THE

PRACTICAL LAND DRAINER

A Treatise on Draining Land.

IN WHICH THE

MOST APPROVED SYSTEMS OF DRAINAGE

AND THE

SCIENTIFIC PRINCIPLES ON WHICH THEY DEPEND, ARE EXPLAINED, AND THEIR COMPARATIVE MERITS DISCUSSED.

WITH

FULL DIRECTIONS FOR CUTTING AND MAKING DRAINS, AND REMARKS UPON THE VARIOUS MATERIALS OF WHICH THEY MAY BE CONSTRUCTED.

Numerously Illustrated.

BY B. MUNN,
LANDSCAPE GARDENER.

NEW YORK:
C. M. SAXTON, BARKER & CO., 25 PARK ROW.
SAN FRANCISCO: H. H. BANCROFT & CO.
1860.

E. O. Jenkins, Printer & Stereotyper,
26 & 28 Frankfort Street.

PREFACE.

UPON few subjects have opposite opinions been advanced with greater confidence, and adhered to with greater pertinacity by those who maintained them, than upon the question of the best method of draining land.

Whilst this may appear extraordinary to those who have but little acquaintance with the subject, it is by no means so to those who have given due consideration to ts theoretical principles, and who have had some practical experience in draining. The causes of these great differences existing between men of unquestionable integrity, talent and experience, may be traced to the diversity of natural causes in operation which render drainage necessary; and in the various methods which have been found by some to succeed, and by others to fail, in situations supposed to be similar, but in reality differing in geological formation, and in external relations.

With the knowledge of these facts, the Author of the following pages has thought he would best serve his readers by placing before them, in something of a systematic form, the principal methods of draining that are considered most desirable; some for their permanent as well as efficient character, others for their less laborious and less costly nature. He has rather sought to explain various systems, and the mode of carrying them out, at the same time pointing out their comparative advantages, than to advance any pet system of his own. He has, consequently, availed himself largely of the labors of other authors of known experience and scientific attainments, and he does not, therefore, claim for himself any originality of principle or practice; yet he hopes that he has explained the principles upon which drainage should be conducted, and how they apply to the various systems for effecting it. And as regards the practical part of the subject, he trusts that the Farmer and the Horticulturist will find information in this volume, in a condensed and useful form, (which they must otherwise seek for in voluminous Agricultural works,) combined with observations arising from some experience in the subject of which it treats.

Everybody knows that water runs down hill; but the Author begs permission to caution the reader against the assumption that, because he knows *that*, he therefore has nothing to learn as regards the preliminary knowledge which is requisite before he begins to drain his land. The want of such knowledge is the too fertile cause of great loss of time, labor and money.

The Author, therefore, strongly recommends the perusal of the few pages of the *Introduction* to those who have not paid attention to the subject which they discuss, before the practical operations of Land Drainage are attempted.

Of the utility of Drainage it is needless in the present day to speak. But could Farmers and Landowners have the benefit of but a tenth part of the experience that has fallen to the lot of the Author, in the course of his professional duties in the improvement of land, he is certain that the extent to which Drainage would be carried would only be limited by the extent of each man's property.

NEW YORK, March, 1855.

NOTE.—The Author may be consulted professionally on the subject of Drainage, by addressing a note to him, Box 3292, Post Office, New York, or to the Publishers, No. 152 Fulton street, New York.

TABLE OF CONTENTS.

INTRODUCTION.

PART I.

CHAPTER I.

CHAPTER II.

CHAPTER III.

CHAPTER IV.

CHAPTER V.

PART II.

CHAPTER VI.

CHAPTER VII.

CHAPTER VIII.

CHAPTER IX.

CHAPTER X.

CHAPTER XI.

CHAPTER XII.

CHAPTER XIII.

INTRODUCTION.

THE PHYSICAL LAWS ON WHICH THE DRAINAGE OF LAND DEPENDS.

THE object sought by the drainage of land, being to remove water from it, it will be well to consider, 1st. *The sources from which the water which we wish to drain away is derived;* and 2d. *The natural laws to which water when still, and when in motion, is subject.*

The operation of heat upon the waters of the ocean and of the land is continuously producing evaporation, by means of which large quantities of water are carried, in the shape of vapor, into the elevated parts of the atmosphere, and are there retained in an invisible form by the agency of electricity. When a change takes place in the electric equilibrium, clouds are formed from the water so raised, which, becoming subject to the laws of physical attraction, are thereby brought in contact with the mountains, and more elevated parts of the earth's surface. Giving out part of their heat, these clouds descend again upon the earth in the form of rain, fog, or snow, according to the temperature, and other meteorological conditions of the atmosphere. Once again upon the surface of the earth, the water becomes subject to its natural laws, and it sinks into the earth, runs down the hill-side, or lays upon the surface, as it may happen from the physical condition of the particular place on which it chances to fall. The water which sinks downward through the soil, and that which is carried through the interstices of rocks and mineral strata, obedient to the laws of gravity and capillary attraction, is distributed beneath the ground to a greater or less depth, until it meets with strata that it cannot penetrate; then it flows along such strata, or accumulates in large bodies, either in hollow basins or diffused through extensive tracts of the subsoil, until it finds vent upon the surface in the shape of springs; or, in other cases, by spreading over a large mass of soil it is held in suspension by it in the same manner as by a sponge. The land so saturated becomes unfit for the purposes of the husbandman.

The natural laws by which water is governed are embraced by the two branches of science, called hydrostatics, which treats of water in a state of rest, and hydraulics, which treats of water when in motion. It is only requisite for the present purpose to state shortly some of the laws of these sciences, without presenting, except so far as is necessary for explanation, the proofs by which the laws themselves are evidenced:

FIRST.—*Water* and *all fluids*, both liquid and gaseous, when *at rest, press equally in all* directions. This results from the extreme minuteness of the particles.

SECOND.—The *pressure* of a column ot *water upon its base depends upon its height and the area of the base;* and not upon the thickness or width of the column.

Suppose *a* and *b*, Fig. 1, to represent two vessels of equal height and capacity. The pressure upon the bottom of *a* will be much less than that on the bottom of *b*. The pressure on *a*, at the bottom,

Fig. 1.

will be equal to a column of water represented by the dotted lines, but the pressure on *b* at the bottom will be equal to a column of water of the size of the whole base of *b*.

The pressure of water in proportion to its height upon a level base is exemplified at *c*, Fig. 1. If a vessel of water be supposed to be divided into four, or any number of equal parts vertically, the pressure upon any part is represented by the sum of the aggregate addition of the parts above. Thus, if the divisions are one foot apart, the pressure at the first division will be equal to a column of water one foot high, at the second division to a column two feet high, and so on.

THIRD.—*Water at rest, and exposed on all parts of its surface to an equal atmospheric pressure, always stands at a uniform level,* whatever be its shape or magnitude. In Fig. 2, the size of the two parts of the vessel are very different; but if water is poured into either end, and the surface of the water is left exposed to the atmosphere, it will rise in both sides to the same level. It is this law (in connection with the law of gravity) that causes rivers to flow and water to percolate through the earth.

From this property of water arises one of the chief causes of springs. It will be an evident consequence of the two first laws above stated, that if water be accumulated in mass within a

Fig. 2.

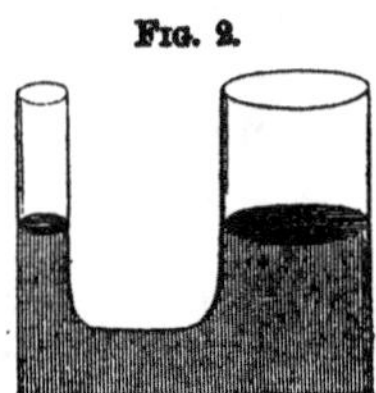

mountain or hill-side, beneath the surface, and it finds egress by a narrow confined passage at a lower elevation, that the force with which it will issue from its lower orifice will be in proportion to the distance it has descended; in other words, in proportion to the pressure of the superincumbent vertical column of water above it; and thence arises the cause of the fountain jet. When, therefore, in draining, a spring is dug into, and the water spouts up with violence, the source of it must be sought for in higher ground, either near or distant. For, in some geological formations the water may travel far beneath the surface.

FOURTH.—*Capillary attraction.* This may be termed a law of the science under consideration, although it is frequently regarded in connection with the general properties of Matter. And it merits attention from the two-fold reason of its apparent contradiction of the last-mentioned law; and also from its being a very constant agent in the production of the evils that it is the object of drainage to counteract. *Capillary attraction is that property of matter which enables water in small tubes or spaces to rise above its common level.* The most simple example is the ascent of water in a sponge. Place a sponge on a glass of water with one end of it just touching the water, and the water will ascend. Take two plates of glass, place their flat sides near each other on one side, and touching each other on the other (Fig. 3), and then put their ends in water. The water will ascend between the plates and stand with a curved surface; the highest part being where the plates touch, as in the shaded line of the figure.

Fig. 3.

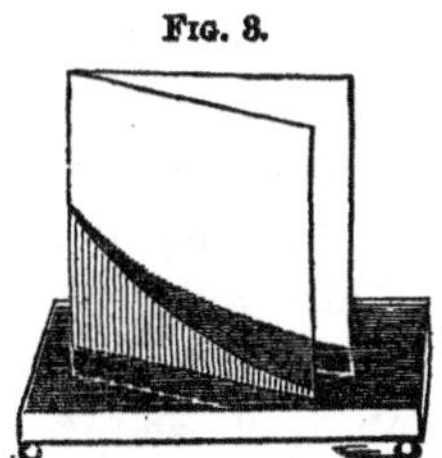

In the same way, as through a sponge, will water ascend, and be held in suspension by the soil, as is familiarly exemplified by the ascent of water in a flower-pot from the saucer beneath it. It will equally ascend into the pot whether the latter contains a plant or earth only.

Capillary attraction, when drainage, whether natural or artificial, is efficient, becomes valuable in its operation upon the growth of vegetation. But it produces results so vastly extensive and so continuous in their effect, that its importance as a cause of surplus water must not be lost sight of by the drainer when he is investigating the condition of land.

Having stated the chief natural laws applicable to the subject, let us briefly refer to their mode of operation.

Whilst upon the earth's surface, the motion of water is regulated by its weight or gravity as the operating force; and that is interfered with principally by the antagonistic force of capillary attraction, by evaporation, and by the conformation of the substances over or through which it passes.

Suppose rain to fall, or water to run down a hillside upon a flat surface below. If that surface be rock, impervious to water, it flows over the surface until it meets with a lower level, by which it passes off; whether that be a fissure in the rock, (in which case it forms a cascade, and afterwards a river in the lower ground beyond,) or by a lower level of equal capacity with that from which it flowed, (in which event it would, as before, flow evenly over it.) But if the rain descend upon soil instead of upon rock, it will be carried by its weight downwards through the soil with greater or less velocity according to the greater or less porous texture of the ground on which it falls. As soon, however, as a portion of the water is *beneath* the surface, capillary attraction begins to exercise a certain amount of force in addition to the one of gravity, which continues its influence. This attraction retains within the pores of the soil (and which may be viewed as an accumulation of minute tubes) a certain portion of the water, and the remainder only then passes downwards by the force of gravity.

It is found, moreover, that the power of capillary attraction varies in different substances; and it varies also in *force*, in a ratio *inverse to the size of the tubes* in which it takes place. Vegetable soils are favorable to the increase of the force in a greater degree than clay, in so far as regards their texture; but they part with water more readily. Whilst clay, from the more minute size of its pores, commands in that respect a greater force of capillary attraction than vegetable soil, and retains water with greater tenacity.

Assume, then, sections to be made on the sides of two hills, one of sand and the other of clay, of equal height, and that rain fell on the top of each, the observer who placed himself at the section to watch the course of the descent of the falling shower through the soil, would find that the water upon each hill-top would sink down perpendicularly a certain distance, and would be attracted, or sucked up, by the top-soil; but if the quantity falling was greater than the rapidity of its downward course, (owing either to the close nature of the soil, or the intervention of rock or other impediment), the surface-soil becoming saturated, the water would be seen to ooze out and run down the face of the sections of each hill. So far, the phenomena presented would be identical. But the following difference in the two sections would be noticed: the perpendicular *distance from the hill-tops*, or *surface-level*, at which the water *first oozed out*, would be found to be much less in measurement in the side of the sandhill, than in the case of the clay-hill; and the reason is this: the superincumbent weight of water requisite to counterbalance the power of capillary attraction in the sandy soil, would be much less than that which was necessary to counterbalance the same power in the close-grained clay; and, consequently (by the natural laws above stated), the column of water above the oozing point would require to be much higher to effect the object. Inasmuch, however, as the force of capillary attraction does not prevent the passage of surplus water through the interstices, in substances upon which it is acting, whilst, on the one hand, it will always retain so much water as its power can command (and the extent of which will, as before mentioned, depend upon the nature of the substance, and the size of the tubes), it, on the other, presents no obstacle to the continued passage of water from the surface to the substrata: the velocity with which it will so pass depending upon various additional considerations.

It is upon the above data that the whole process of *Under-draining* is based.

But we must now inquire what becomes of that part of the water which is not held in suspension near the surface by capillary attraction, when, by its gravity, it has accumulated in quantities in the substrata below. The passage downward continues in perpendicular lines from the surface, until its further progress in that direction is impeded (as has been before observed) by some non-porous obstacle, when, if it cannot find vent laterally, it accumulates, and, obeying the law of finding a level, it forms a "water-line" at a given depth (sometimes called a *water-table*), which level is elevated nearer to the surface, in proportion to the quantity falling upon such surface. Thus it remains; stagnant, except so far as its mass is diminished by the capillary attraction* constantly going on in the soil above the water-line, which is induced to supply the loss of moisture in it, occasioned by the evaporation which is constantly emanating from the surface.

In this condition it is that, for profitable farming, land, when this water-line is so near the surface as to

* Professor Leslie calculates the rise of water in coarse sand or loam as follows: If the gravel were divided into spaces of a hundredth of an inch, the water would ascend 4. inches, and so on to a ten-thousandth part of an inch, when the height would be 25. feet. To make this calculation good, the pores should be continuous, and always open. In most soils, the cohesion of the particles, and consequently the diameter of the pores, is affected by the contact of water. In chalk this is not the case. To speak first of this substance: In chalk, the pores, which are not to be discovered by a strong microscope, are always open. A piece of dry chalk, six inches high, of which the foot is placed in contact with the surface of the water, absorbs about one-third of its bulk, and about one-fourth of its weight; it will become quite saturated in a short time, say one hour, so that it will take up no more, though immersed in water. Moreover, if suddenly immersed when dry, it will not be thoroughly saturated, a portion of the air contained in the pores being impounded in the centre. If a second piece be placed on that saturated, the water will pass through one to the other, and so on. About thirty pieces, piled on each other in a glass tube, the point of contact between some of them being not more than the surface of a pin's head, become saturated in about two months, so that the top piece, when immersed, would not take up more than the turn of the scale in addition. Hence, we see how water is lifted in chalk, which will not part with a single drop by drainage. Clay, when submitted to water, also becomes saturated by capillary attraction, but much less rapidly than chalk.—*Gardener's Chronicle*, 23*d July* 1853.

interfere with vegetation, requires to be drained, no less than it does when the water stands on the surface; and this introduces to us the question of the manner of effecting that object.

PART I.

PRINCIPLES AND SYSTEMS OF DRAINAGE.

CHAPTER I.

THE EXAMINATION OF LAND REQUIRING DRAINAGE.

THE Art of Drainage may be defined to be, that preparation of land for the purpose of the husbandman which places it in a fit condition for retaining so much and no more moisture, and for such periods of time, as is best adapted to the vegetation and growth of his crops. This, of course, includes the removal from the land of superabundant water, whether upon the surface or beneath it. No one will assume that it is practicable in all cases to attain perfection in this respect; but in this, as in all other human pursuits, he who would succeed must present the *ideal* at least of perfection to his mental vision, as the object to be aimed at; and then he must use the best means within control to attain as nearly to the image set before him as circumstances will permit; always, in his efforts, keeping in mind the fact that his personal energies to attain his object, being within his own control, must be lavished without stint. Much has been done in many matters with small means, but with *such* energy.

It should be observed, moreover, that placing the soil in a fit condition as to the quantum of moisture for vegetation, is not the sole object of moment to the husbandman to be attained by draining. Another and very important consequence that results from it is, the elevation of the temperature of the soil immediately under the surface. This, during the greater part of the year, is of much importance to vegetation.

The advantages of draining land are now so generally known, that it would be superfluous to do more than advert to them. Suffice it to say, that, whilst it renders much waste land valuable which was before useless, it renders doubly productive much land that already is valuable.

In the report on draining, in the Transactions of the N. Y. State Agricultural Society, for 1848, the committee state, in reference to *upland*, "that there is not *one farm* out of every *seventy-five* in this State but *needs draining*—yes, much draining—to bring them into high cultivation; nay, we may venture to say, that every wheat-field would produce a larger and finer crop if properly drained."

In order to simplify the subject, it is proposed to divide this volume into two parts.

The FIRST PART will treat of the *principles* of Drainage, the various *Systems* in use, and their adaptation to different descriptions of land.

The SECOND PART will contain *practical directions* for the construction of Drainage, and will also explain the *materials* that can be made available for the purpose.

The first object to be looked at in regard to draining, *is the source and cause of the superabundant water* in the land about to be drained. This must be sought for in cases where it is not self-evident by an examination of the soil and subsoil; and of the geological formation of the situation, and of the surrounding country; except in cases where the source or cause of the wet condition of the land is apparent. The examination of the soil and the subsoil is made by digging large holes down several feet deep, at distances around the land, or by a trench dug along the sides, (which may afterwards form part of the drains,) taking care to examine especially such parts as either in reference to the quantity of water present, or the conformation of the surface, appear to indicate the probability of variation in the substrata. This examination should be made to a depth of five feet at least, if the nature of the ground permit; but if the surface soil is a shallow layer upon rock, the nature of the rock should be examined, and its character, whether of porosity or of impermeability, ascertained. This examination as to the soil, if deep, will enable the drainer to arrive at some approximate idea of the depth at which the water is impeded, and the course of its flow beneath the surface, if it be not stagnant.

The geological examination is, however, by no means easy even to the experienced eye. The distribution of soils upon the surface of the earth, is very various as to quality, texture, and thickness. The formation of the subjacent rocks on which they rest, is no less so. The direction of the rocky strata as to their inclination, or "dip," as it is called, is also changeable and endless. Some strata are porous, readily admitting the passage of water; others in a much less degree so; and again others, whilst they are impervious to water, as regards their solidity, are filled, nevertheless, with fissures and cracks, which permit the free ingress and egress of the water through them. It will be obvious, therefore, that the great secret which it is desirable to the drainer to ascertain, is the effect of these numerous, and often conjointly operating causes, upon the course of the water with which he has to deal.

In level districts of country, the examination of the surface and the subsoil is usually sufficient to direct to tolerably satisfactory conclusions for practical purposes; but in hilly and mountainous districts, it will be evident that the underground aqueous currents must be dependent upon a multiplicity of circumstances, the effects of which can only to a limited extent be understood, whilst still such of them as can be unravelled, afford valuable information to guide the operations of drainage.

In order to give some general ideas on the subject, the following extract from a valuable article on Drainage is given, which will afford data upon which to found more particular observations of different localities:

"*Should a mountain consist of concentric layers of different rocks arranged mantle-shaped around it,* then water will descend between the lines of junction of the rocks; and should the masses or beds of rock be of different extents, and thickness, and consistence, then the water will either appear at the surface of the ground as a spring, from the subjacent rock of a close texture, or it will descend yet lower, and be absorbed by the subjacent rock of a porous texture. In this manner the harder rocks cause the springs to appear at a high elevation, while the porous ones convey the water to a lower level, until it meets with a resisting substance to cause it to come to the day. In any case the farmer cannot do any thing until the water indicates its presence on the surface of the ground, either at a high or low elevation; and then he should take measures accordingly to remove it.

"To illustrate the cases now alluded to, suppose

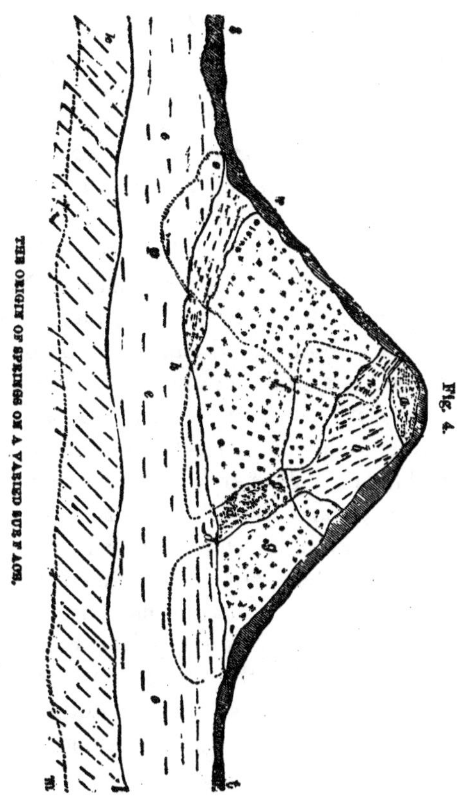

Fig. 4 to represent a hill composed of different rocks of different consistence. Suppose the nucleus rock *a* to be of close texture, when the rain falls upon the summit of the hill, which is supposed not to be covered with impervious clay, but with vegetable mould, the rain will not be absorbed by *a*, but will pass down by gravity between *a* and *b*, another kind of rock of close texture. When the rain falls in greater quantity than will pass *between* these rocks, it will overflow the upper edge of *b*, and pass over its surface down to *c*; but as *c* is a continuation of the nucleus impervious rock *a*, a large spring will flow down the side of the hill from *c*, and render the ground quite wet to *d*, where, meeting another large stratum of impervious rock, it will burst out to-day a large spring at *d*, which will be powerful in proportion to the quantity of rain that falls on the mountain. On flowing down *b*, part of the water will be intercepted by the rocks *f* and *g*, both of which, being porous, will absorb and retain it until surcharged. The surplus water meeting with the impervious rock *e*, will be partly thrust out to-day along the black line *d h*, on the one hand, and *d i* on the other, when the whole line *h i* will present a long dark line of wet oozing out of the soil, with the spring *d* in the centre, and which darkness and dampness will extend down the inclined ground as far as the upper line *k l* of that porous stratum of rock. Part of the water absorbed by the porous rocks *f* and *g* will be conveyed under the impervious rock *e*, and come out at their lowest extremities, following the curved dotted lines *h d* and *d i*, and continue to flow on until it reaches the lowest extremity of *e* in the dotted line *k l*, where it will be absorbed by the porous rock *m*.

"By such an arrangement of rocky strata on the side of a mountain range, will be exhibited specimens of both wetness and dryness of soils. The summit *a* will be wet, and so will the surface of *b*, but the surfaces of *f* and *g* will be dry. Again, the surface of *e* will also be wet, but less so than that of *b*, because part of the water is conveyed by *f* and *g* under *e* to the dry stratum *k l*, which being probably thicker, and, at all events, of greater extent, will be drier than either *f* or *g*. On another side of the surface of the hill another result will take effect. The rain falling on the summit *a* will descend between *a* and *n*, as far as the lowest extremity of *n* along the dotted line *o p*, which being under the impervious rock *e*, the water will continue to flow out of sight until it descends to *k l*, where it will be absorbed by the porous rock *m*, and thus never appear at all either as a spring or a line of dampness. But should the quantity of rain at any time be greater than what will pass between *a* and *n*, it will overflow *n* and be absorbed in its descent by the porous rock *f*, which, after becoming surcharged, will let loose the superfluous water in the line *h r*, upon the continuation of the rock *n*, part of which will come to-day along the line *h o* of the imperious rock *e*, and part conveyed down by *o p* to the porous rock *k l*, where it will be absorbed. Thus, on this side of the hill, as long as little rain falls, none but its summit will be wet, and all the rest will be dry, though the surfaces of *f* and *k* will always be drier than those of *n* or *e*; but after heavy rains dampness will show itself along the line *h r*, and will extend itself even to the line of *k l*, should the rain continue to fall some time.

"The line *s* by the summit *a* to *t* is the mould line pervious to moisture, and which is here represented as is frequently exhibited in nature, namely, a thickness of soil on the southern side of the hill, as from *a* to *i*, and a thickness of soil on the northern basis, as from *r* to *s*; but a thinness of soil on the southern face, as from *a* to *r*. It is not pretended that this figure is a truly geological portrait of any mountain. But such overlying and disconnected strata do occur over extended districts of hilly country which produce springs much in the way described. Similar courses of water occur in less elevated districts, though it remains more hidden under the deeper alluvial rocks."*

From the foregoing diagram explanatory of the geological distribution of soils, and of the strata upon which they rest, it will be evident that if a section be cut through ground which slopes, it will frequently happen that the substrata lying more or less horizontally will "crop out," or strike the slope as it is ascended, at right angles, or nearly so, as is shown in the figure.

Fig. 5.

THE USUAL POSITION OF SUBSTRATA IN REFERENCE TO THE SURFACE SOIL.

Bearing in mind the preceding remarks upon the passage of water through strata of various degrees of permeability, it will be obvious that when water from the top of the slope has sunk through such of the upper strata as are porous, and in its descent comes to strata that, from their closer texture, oppose its downward course, it will flow along such strata until the body of the latter becomes saturated, and the

* Stephens.

surplus water will then come out upon the slope and run down its face.

It will be perceived, from the brief view which has been taken, that the principal causes of mischief to undrained land arise from one or more of the following conditions:

1. Where water has accumulated beneath the surface, and originated springs;
2. Where, from the close nature of the substrata, it cannot pass freely downward, but accumulates, and forms its level, or water-line, at a short distance below the surface; and,
3. Where, from the clayey or close texture of the soil, it lies on the surface, and becomes stagnant.

The preliminary examination of the ground above directed, is made with the view to ascertain as nearly as possible whether the water to be got rid of proceeds from one or more, and which, of the above causes.

Swamps, bogs, and morasses, generally are occasioned by the first and second of the above conditions; and that surface-wet state, that often prevails in tenacious soils, that have a comparatively even level appearance, usually arises from the last cause named above.

CHAPTER II.

THE VARIOUS SYSTEMS OF DRAINAGE.

UPON the cause which is found (from the examination of the land recommended in the preceding chapter) to produce the water that is to be drained away, will depend the system of drainage that it is most expedient to adopt, in order to remove it.

There are three systems of drainage which have been most in use. These are, Surface Draining, Frequent or Thorough Draining, and Deep Draining.

Surface Draining consists in cutting channels for water to pass through upon the surface, which are left as permanent open ditches.

Frequent, or Thorough Draining, or *Under-draining*, as it is often called, consists in opening numerous drains at various depths near the surface, or not below three or four feet from it, and filling them again, after providing a duct for the passage of water at the bottom of them, either open or formed of porous materials.

Deep Draining consists in making drains in the same manner as for Frequent or Thorough Drains, but much fewer in number, and placed at greater depths—from four or five to eight feet, or even more, according to the nature of the ground and the formation of its surface line. This is called, also, "Elkington's System," from the name of its originator.

For the purpose of *intercepting springs and deeply-seated* masses of *water*, the value of *deep draining* is generally acknowledged; but its efficiency, as compared with the system called Thorough Drainage, for the purpose of removing water present, or falling upon the surface in the shape of rain, in subsoils of a stiff, clayey character, and for keeping such soils in a favorable state for agricultural purposes, has been made matter of great dispute with professional drainers of eminence.

It may serve to elucidate the question of the comparative advantages of the two systems, to state in as few words as possible the principal features in the dispute between the advocates of each.

The *deep drainers* contend that, as water always seeks the lowest level, the deeper the drain is placed, the lower must be the under-ground *water-line*, or level of surplus water; because, the water that passes down from the surface to the depth of the drain, can only accumulate below it; for, as soon as it has risen below it to the level of the bottom of the drain, it will then enter the drain *at the bottom*, through its porous material, and be carried away by it. Consequently, they say, the deeper the drain, the greater will be the mass of thoroughly-drained soil above it.

This is undoubtedly correct, provided the deep drains be near enough together, and the soil above them be porous enough to allow of the passage of water through *with sufficient rapidity* to keep the surface usually in a fit state for agriculture. In many soils and situations, it may be the case. But if the subsoil be of a stiff, clayey, or very tenacious nature, the *shallow drainers* contend that thorough or frequent drainage becomes the most efficient; because, they conceive that the water cannot descend far below the surface in clay; and that it is either held, therefore, in suspension by the soil, or, if saturated, that it remains on the surface, and that it does not pass through to deep drains below. And this may be so in very stiff soils, so far, at least, as to render it impracticable, (*looking at the quantity of falling water to be got rid of in a given time*,) to prevent the water-line rising nearer to the surface than is desirable, from the slow rate at which, in such soils, water percolates downwards. But the idea that water will not *slowly* continue its passage down through clay, is erroneous. *The principle* of deep draining is, therefore, not really the subject matter in the dispute, so much as its expediency and practical utility, as compared with more shallow and numerous drains, in some soils and situations.

That this is the solution of these different views of the matter, appears to be the fact, from the circumstance, that in many instances in which deep draining has been adopted, after shallow drains had been found ineffectual *upon the same* piece of land, (and in which, consequently, the two sets of shallow and deep drains have existed together,) it has been ascertained, that after heavy storms, the deep drains have given out water long after the shallow ones ceased to do so. It is clear, therefore, that had the deep drains been *absent*, the shallow drains could only have taken the water away which was above, or on the same level with themselves.

But it has been wrongly called by the advocates of either side a dispute about deep draining, and frequent or thorough draining: for as Elkington's deep drainage provides for the free passage of water by open subterranean ducts, and as Smith, of Deanston, (the great promoter of the system known as thorough draining,) directs especially the placing of an impervious material over the drains to prevent the downward passage of water directly into them from the surface, the *mode of operation* of those drains is the same as that of drains cut deeper and at greater distances apart; whatever may be the relative advantages of the two systems as to their efficiency.

The dispute in reality has been, *What is the true way* in which the drains act? And the questions put in issue by the parties may be stated thus: Whether deep drains on the Elkington system are applicable to all cases; those arising from the pressure of springs or stagnant water, and those arising from surplus surface or rain water, in stiff soils and subsoils? And whether, assuming that frequent drainage is a preferable system for the latter class of cases, this frequent drainage should be at a depth of three feet as a *maximum*, as advocated by the shallow drainers, or that four feet should be a *minimum* depth, as advocated by the deep drainers? And further: What should be the distance apart of the drains? These are the questions now remaining at issue between the most eminent drainers in England.

In some situations, where, for instance, the examination of the land showed that springs from high ground were partly the cause of the evil, and the nature of the substrata another, it would be right to cut off the springs upon Elkington's system, and then drain the land generally upon the frequent drainage principle.

The system of surface drainage, by open drains, requires no particular notice in this chapter, as its mode of action is self-evident.

We will now proceed to explain the above systems, and the method of carrying them out.

CHAPTER III.

DEEP DRAINAGE.

THE system of *deep drainage*, is called *Elkington's method*, having originated with Mr. Joseph Elkington, a farmer, who resided in Warwickshire, England, and was first practiced by him about 1764. His fields being very wet, and rotting many of his sheep, he dug a trench four or five feet deep, with the view of discovering the cause of the wetness. Having a suspicion that the drain was not deep enough, he forced an iron crowbar four feet below the bottom of the trench, and, on pulling it out, a great quantity of water welled up through the hole it made. He was led to infer from this, that large bodies of water are pent up in the earth, and constantly injuring the surface-soil, but which may be let off by tapping with an auger or rod. This discovery introduced a complete revolution in the art of draining.

The *principles* of Elkington's mode of draining seem to depend on these three alleged facts: 1. That *water from springs is the principal cause of the wetness of land*, which, if not removed, nothing effectual in draining can be accomplished. 2. *That the bearings of springs to one another must be ascertained* before it can be determined where the lines of drains should be opened; and by the bearings of springs, is meant that line which would pass through the seats of *true* springs in any given locality. *Springs* are characterized as *true* which continue to flow and retain their places at all seasons; and *temporary springs* consist of bursts of water, occasioned either by heavy rains, causing it to appear on the surface sooner, or at a *higher* level than permanent springs, or by true springs leaking water, and causing it to appea. on the surface at a *lower* level than themselves; and, if such springs are weak, their leakage may be mistaken for themselves. It is evident, that if drains are formed through these *bursts* of water, no effectual draining takes place, which can only be accomplished by the drain passing through the line of true springs. 3. *That tapping the spring* with the auger *is a necessary* expedient, *when the drain cannot be cut deep enough* to intercept it.*

The causes of the wetness of land, to remedy which this system was put forward, were primarily those arising from springs, and oozings of water issuing from elevated and hilly districts of ground, the characters of which frequently prove the means of rendering the grounds below wet and swampy; for the general moisture of the atmosphere being condensed in much greater quantities in such elevated situations, the water thus formed, as well as that which falls in rain, and sinks through the superficial porous materials, readily insinuates itself, and thus passes along between the first and second, and still more inferior strata, which compose the sides of such elevations, until its descent is retarded, or totally obstructed by some impenetrable

* Johnston on Elkington's Mode of Draining.

substance, such as clay. It there becomes dammed up, and ultimately forced to filtrate slowly over it, or to rise to some part of the surface, and constitute, according to the particular circumstances of the case, different watery appearances in the grounds below. These appearances are oozing springs, bogs, swamps, or morasses, weeping rocks (from water slowly issuing in various places), or a large spring or rivulet, from the union of small currents beneath the ground. This is obvious from the sudden disappearance of moisture on some parts of lands, while it stagnates, or remains till removed by the effects of evaporation, on others; as well as from the force of springs being stronger in wet than in dry weather; breaking out frequently after the land has been impregnated with much moisture in higher situations, and as the season becomes drier ceasing to flow, except at the lowest outlets. The force of springs, or proportion of water which they send forth, depends, likewise, in a great measure, on the extent of the high ground on which the moisture is received and detained, furnishing extensive reservoirs, or collections of water, by which they become more amply and regularly supplied. On this account, what are termed bog springs, or such as rise in valleys and low grounds, are considerably stronger, and more regular in their discharge, than such as burst forth on the more elevated situations on the sides of eminences.*

In order to remove the evil consequences referred to, *caused by water passing between the porous and impervious strata* of mountains and hills, and producing springs, the mode of draining is that of intercepting

* Johnston on Draining.

the descent of the water or spring, and thereby totally removing the cause of wetness. This may be done where the depth of the superficial strata, and consequently of the spring, is not great, by making hori-

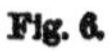
Fig. 6.

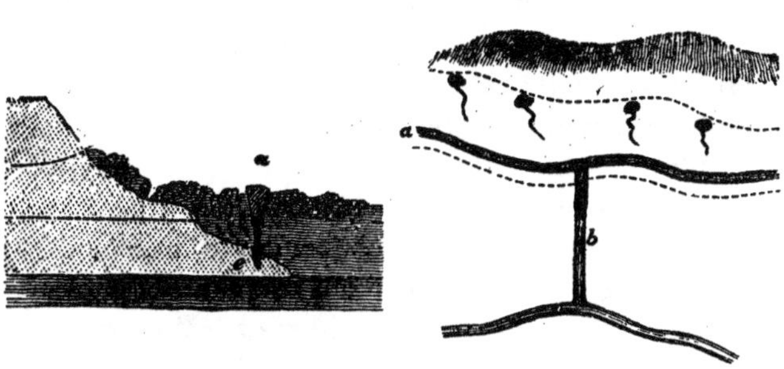

zontal drains (Fig. 6, *a*) of considerable length across the declivities of the hills, about where the low grounds of the valley begin to form, and connecting these with others, *b*, made for the purpose of conveying the water thus collected into the brooks or rivulets, *c*, that may be near. Where the spring has naturally found itself an outlet, it may frequently only be necessary to bore into it, *e*, or render it larger and of more depth; which, by affording the water a more free and open passage, may evacuate and bring it off more quickly, or suit it to a level so greatly below that of the surface of the soil, as to prevent it from flowing into or over it.

Where the *uppermost stratum* is so extremely thick as not to be easily penetrated, or where the springs formed by the water passing from the higher grounds may be confined beneath the third or fourth strata of the materials that form the declivities of hills or elevated grounds, and by this means lie too deep to be penetrated by the cutting of a ditch, or even by boring, the common mode of cutting a great number of drains to the depth of five, six, or more feet across the wet morassy grounds, and afterwards covering them in such a manner as that the water may suffer no interruption in passing away through them, may be practiced with advantage; as much of the prejudicial excess of moisture may by this means be collected and carried away, though not so completely as by fully cutting off the spring.*

In the drainage of *wet or boggy grounds arising from springs of water beneath them*, a great variety of circumstances are necessary to be kept in view. Wet grounds of these kinds may be arranged under three distinct heads: *First*, such as may be readily known by *the*

Fig. 7.

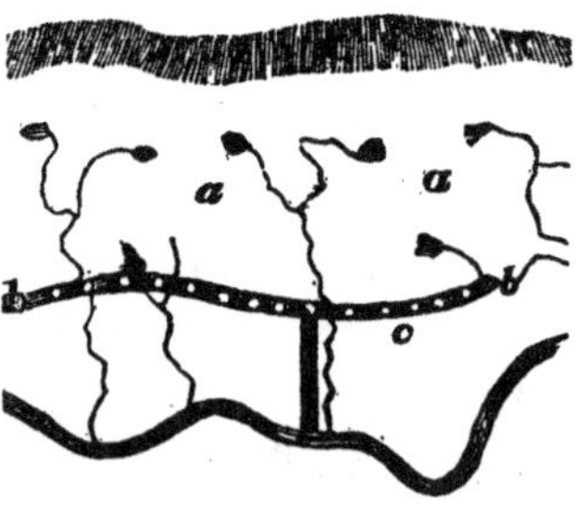

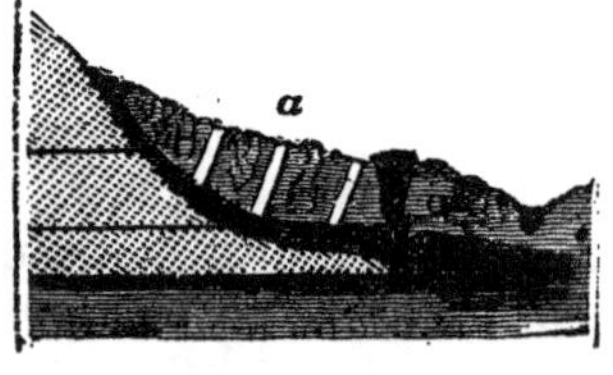

springs rising out of the adjacent more elevated ground in *an exact or regular* line along the higher side of the wet surface. *Secondly*, those in which the *numerous*

* Loudon.

springs that show themselves are *not kept* to an exact or *regular line* of direction along the higher or more elevated parts of the land, but *break forth promiscuously* throughout the whole surface, and particularly towards the inferior parts, (Fig. 7, *a*,) constituting shaking quags in every direction that have an elastic feel under the feet, on which the lightest animals can scarcely tread without danger, and which, for the most part, show themselves by the luxuriance and verdure of the grass about them; and, *Thirdly*, that sort of *wet land from the oozing of springs*, which is neither of such great extent, nor, in the nature of the soil, so *peaty* as the other two, and to which the term *bog* cannot be strictly applied, but which, in respect to the modes of draining, is the same.*

In order to direct the proper mode of cutting the drains, or trenches, in draining lands of this sort, it will be necessary for the draining engineer to make himself perfectly acquainted with the nature and disposition of the strata composing the higher grounds, and the connection which they have with that which is to be rendered dry. This may, in general, be accomplished by means of levelling, and by inspecting the beds of rivers, the edges of banks that have been wrought through, and such pits and quarries as may have been dug near to the land. Rushes, alder-bushes, and other coarse aquatic plants, may also, in some instances, serve as guides in this business; but they should not be too implicitly depended on, as they may be caused by the stagnation of rain-water upon the surface, without any spring being present. The line

* Johnston on Draining.

of springs being ascertained, and also some knowledge of the substrata being acquired, a line of drain (Fig. 7, *b b*) should be marked out above or below them, according to the nature of the strata, and excavated to such a depth as will intercept the water in the porous strata before it rises to the surface. The effect of such

Fig. 8.

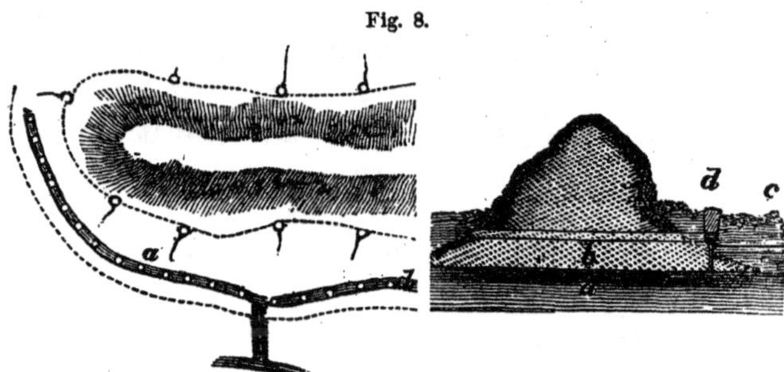

drain will often be greatly heightened by boring holes in their bottom with the auger. Where the *impervious* stratum (Fig. 8, *a*) that lies immediately beneath the porous, *b*, *has a slanting direction* through a hill o· rising bank, the surface of the low lands will, in general, be spongy, wet, and covered with rushes on every side, *c*. In this case, a ditch or drain, *d*, properly cut on one side of the hill or rising ground, may remove the wetness from both. But where the impervious stratum dips, or declines more to one side of the hill or elevation than the other, the water will be directed to the more depressed side of that stratum; the effect of which will be that one side of such rising ground, will be wet and spongy, while the other is quite free from wetness.*

* Loudon on Draining.

In cases *where the banks or rising grounds are formed in an irregular manner* (Fig. 9), and from the nature of the situation, or the force of the water underneath, springs abound round the bases of the protuberances, the ditches made for the purpose of draining should always be carried up to a much higher level in the side of the elevated ground than that in which the water or wetness appears, as far even as to the firm, unchanged land. By this means, the water of the spring may be cut off, and the ground completely drained; which would not be the case if the trench or drain were formed on the line of the loose materials lower down,

Fig. 9.

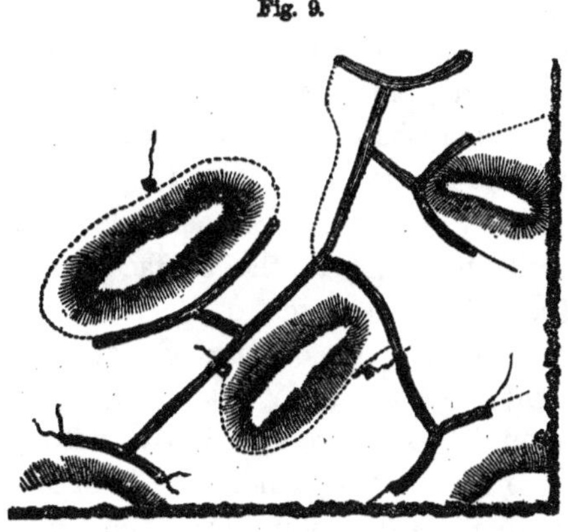

where the water oozes out; which is liable to mislead the operator in forming the conducting trench, or that which is to convey the water from the cross-drain on the level of the spring to the outlet or opening by which it is discharged (Fig. 9). But where the main or principal spring comes out of a perpendicular or very steep bank, at a great height above the level of the outlet into which it may discharge itself by means of a drain, it will neither be necessary, nor of any utility, to form a deep trench, or make a covered drain all the way from such outlet up to it; as, from the steepness of the descent, the water would be liable, when the drain was thus cut from the thin strata of sand and other loose materials always found in such cases, to insinuate itself under the bricks, stones, or other substances of which the drain was formed, to undermine and force them up by the strength of the current; or, probably, in some instances, block the drain up by the loose sand, or other matters which may be forced away or carried down by it. In situations of this kind, Johnston observes, it is always the best way to begin just so far down the bank or declivity as, by cutting in a level, the drain may be six or seven feet below the level of the spring; or of such a depth as may be requisite to bring down the water to a level suitable to convey it away without its rising to the surface, and injuring the lands around it. The rest of the drain, whether it be made in a straight or oblique direction, need not be deep; and may, in many instances, be left quite open; it should, however, be carefully secured from the treading of cattle, and, where the land is under an arable system of cultivation, also from the plough. Where it is covered, the depth of about two feet may be sufficient. There will not, in such drains, be any necessity for the use of the auger in any part of them.*

In the cases of *hills covered with stiff clay* and tenacious soils, where the large quantity of water falling on

* Loudon.

their surface cannot sink into them, but flows down over the surface, the mode of draining them is shown by reference to Fig. 10.

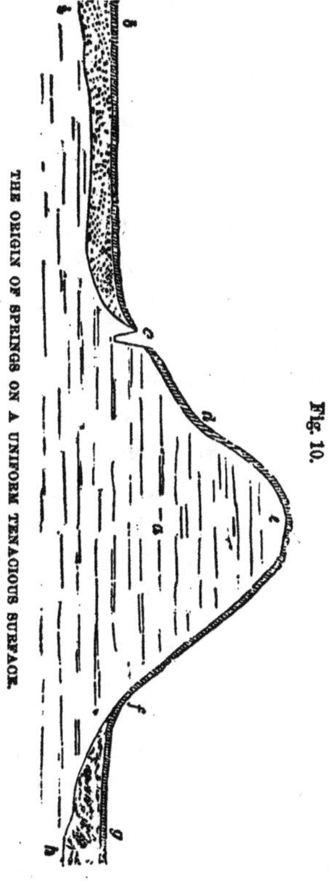

Fig. 10.

THE ORIGIN OF SPRINGS ON A UNIFORM TENACIOUS SURFACE.

The hill in Fig. 10 being supposed to be covered saddle-shaped with an impervious stratum of clay, no water can descend *into* it, but will flow *over* it: *a* is the clay stratum; *b* also an impervious stratum, but not so much so as *a*, containing veins of sand and nodules of stones. It is clear that the whole extent of ground from *e* to *b* will be wet on the surface, and the wetness will not exhibit itself in bands, but be diffused in a uniform manner over the whole surface; but as *b*, in this case, is not so tenacious as *a*, the side of the hill from *e* to *c* will always be wetter than the flat ground from *c* to *b*, because some of the water will be absorbed and kept out of sight in the looser clay *b*. The only method of intercepting the large body of water in its descent down *d* is to cut the deep drain at *c*, not only sufficiently large to contain all the water that may be supplied from above *c*, but so deep as to catch any oozing of water from *a* toward *b*. What the depth of this drain should be, it is not easy to determine without farther investigation, and to enable that investigation to be made, a large drain should be cut on the flat ground in the line from *b* to *c*, which will also answer the purpose of leading away the water that will be collected by the transverse drain *c*. Suppose the subsoil from *b* to *i* is four feet thick, then this leading drain should be made one-half foot deeper, namely, four and a half feet, in order that its sole may be placed in impervious matter; and in this case, the drain *c*, of the depth of six feet, may suffice to keep the flat ground dry. But if from *b* to *i* is eight or ten feet in depth, then it would be advisable to make the leading drain from *b* to *c* at least six feet deep, in order to drain a large extent of ground on each side of it, and the drain *c* may still do at its former depth, namely, six feet. Should the bottom of the leading drain get softer and wetter as the cutting descends, its depth should either be carried down to the solid clay at *i*, or perhaps it would be well to try auger holes in the bottom, with the view of ascertaining whether the subjacent water might not rise to and flow along it. The expedient of boring will be absolutely necessary, if the depth from *b* to *i* decreases as the distance from the hill increases, for there would be no other way of letting off the water from the basin of the clay from *i* to *c*. Should the flat ground be of considerable extent, or should the face of the plain undulate considerably from right to left, a leading drain will be required in every hollow; and each of them should be made deeper or shallower, according as the subsoil is of an open texture or otherwise, bearing in mind that the bottom or *sole* of the drain should, if possible, rest upon an impervious substance, otherwise the water will escape through the pervious matter, and do mischief at a lower level. The subsoil between *g* and *h*, being supposed to be gravel, or other porous substance, it is clear that no drain is required at *f* to protect the soil between *f* and *g*, as the porous subsoil will absorb all the water as it descends from *e* to *f*.

As to the wet surface of the hill itself *c d e f*, it being composed of impervious clay, must be dried on the principle of surface-draining; that is, if the ground is in permanent pasture, a number of transverse open* sheep drains should be made across the face of the hill, and the water from them conveyed in open ditches into the great drain *c*; or if the ground is under the plough, small covered drains will answer the purpose best; and the contents of these can be emptied into the large drain *c*, and conveyed down the large leading drain to *b*. Thus in Fig. 11, *a b* is the main drain along the flat ground into which the large drain *c b* and *d b* flow. It may be observed here that when one *large* drain enters another, the line of junction should not be at right angles, but with an acute angle in the line of the flow of water, as at *b*. The open surface-drains in permanent pasture exhibit the form as represented in this figure, where the leaders *e f* and *g h* are cut with a greater or less slope down the hill according to the steepness of the acclivity, and the feeders across its face nearly in parallel rows, into their respective leaders. In this way the water is entirely intercepted in its descent down the hill. Where small drains enter larger, they should not only enter with an inclination, as remarked above, but where they come from opposite sides,

* See Chap. on different kinds of drains.

as in this case, they should enter at alternate distances, as seen in the case of the three drains above *f*, and not as shown in the fourth and fifth drains. The large

Fig. 11.

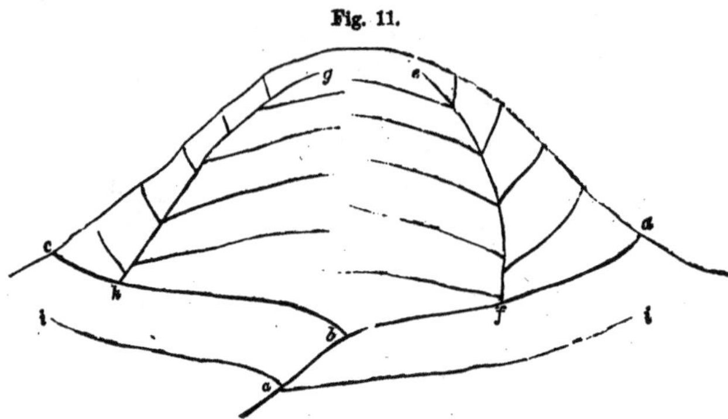

A PLAN OF SHEEP DRAINS ON A HILL OF IMPERVIOUS SUBSOIL.

drain *c b d* may either be left open or covered. Should it form the line of separation between arable ground and permanent pasture, it may be left open, and serve to form a fence to the hill-pasture; but should the entire rising ground be under the plough, this, as also the main drain *a b*, and all the small drains, should be covered.*

Where springs, or oozings of water, rise around gravelly eminences, standing isolated upon a bed of clay, or other impervious matter, a circumvallation of drain around the base of the eminence, begun in the porous, and carried into the impervious substance, having a depth of perhaps from five to seven feet, and connected with a main drain along the lowest quarter of the field, will

* Stephens.

effectually dry all the part of it that was made wet by the springs or oozings.

Bogs and *marshes* have been drained with great effect by Elkington's method, which rested on basin-shaped hollows in clay; and, when this is of considerable depth, the only way of draining them is by bringing up a deep cut from the lowest ground, and passing it through the dam-like barrier of clay. But it not unfrequently happens, that gravel or sand is found at no great depth below the clay on which bogs rest; in which case, the most ready and economical plan, is to bore a hole or holes, in the first instance, through the clay, with an auger five inches in diameter, and, after the water has almost subsided, to finish the work by sinking wells through the clay, and filling them up with small stones, to within two feet of the top.*

* Stephens.

CHAPTER IV.

FREQUENT OR THOROUGH DRAINAGE.

FREQUENT or Thorough Drainage is effected by means of small drains, placed under ground, at short distances from one another, and into which the water, as it falls upon the surface, or already present, finds its way through the porous soil, and by which it is conveyed by main drains, made in connection with the small, to an outlet.

This system of drainage was, about the year 1832, brought prominently into notice by James Smith, Esq., of Deanston, Sterlingshire, in Scotland, as providing frequent opportunities, both for the water rising from below, as well as for that falling on the surface, to pass freely and completely off; and hence the name of Frequent Draining, which has been given it. Mr. Smith states that in Scotland, much more injury arises to land there from the retention of rain-water, than from springs; and he, by implication, treats Elkington's Deep Drains as efficient for remedying wetness arising from the latter, but the Thorough Drainage as especially applicable to the former, and as effective for both, under ordinary circumstances.

The principal things for consideration, in order to carry out efficiently the system of frequent or thorough drainage of land, are:—

FIRST.—The *situation of the drains* and *their fall*, or the inclination at which they are to be made to ensure the ready passage of water through them.

SECOND.—Their *size* and *depth*.

THIRD.—Their *distance apart*.

As to the *situation* of the drains, the small drains are distributed at intervals over the whole surface of that part of the ground which requires drainage, and the main drains into which they empty themselves are placed on the lowest part of the ground.

The first, and a most important question in placing drains, is the *direction in which they should run* with regard to the inclination of the surface of the ground. This subject is another "vexed question" between experienced drainers; but certainly in this instance the one side of the argument has undoubtedly the advantage. This question has been, whether the small drains which first collect and receive the water from the land should be placed *across* the slope of the land, or *down* the slope. The latter is the proper direction. Because in cutting drains across a sloping surface, in cases where the different substrata "crop out" upon the slope (see Fig. 5, page 28,) unless they are put in at the precise point where the substrata so crop out, (and these are very irregular in point of thickness), they may, in a great measure, prove nugatory. For although one drain is neaı another, from the rise of the ground none of them may reach the point sought, whereas in carrying a drain up the direction of a slope, it is impossible to miss the extremity of every substratum passed through.

This view of the direction of drains is supported by Mr. Smith, who says, "drains drawn across a steep, cut the strata or layers of subsoil transversely; and, as the stratification generally lies in sheets at an angle to the surface, the water passing in or between the strata, immediately below the bottom of one drain, nearly comes to the surface before reaching the next lower drain. But as water seeks the lowest level in all directions, if the strata be cut longitudinally by a drain directed down the steeps, the bottom of which cuts each stratum to the same distance from the surface, the water will flow into the drain at the intersecting point of each sheet or layer, on a level with the bottom of the drain, leaving one uniform depth of dry soil."*

And the accuracy of Mr. Smith's observations is demonstrated by Mr. Stephens in the following illustrations and remarks:

"Without taking any other element at present into the argument than the single proposition in hydraulics that water seeks the lowest level in all directions, I shall prove the accuracy of Mr. Smith's conclusions by simply referring to Fig. 12, which represents a part of a field all having the same, and that a steep, declivity, and which is laid off in the ridges *a b c d e f*, up and down the slope; but the three ridges *a b c* have drains across them, and the other three ridges have drains parallel with them, the oblique drains being made at the same distance from each other as the up and down ones, whatever that distance may be. Now, when rain falls on and is absorbed by the ridges *a b c d e f*, it will naturally make its way to the lowest level, that is, to the bottom of the drains; and, as the ground has the

* Smith on Thorough Draining.

same declivity, the water will descend according to the circumstances which are presented to it by the positions of the respective systems of drains. On the ridges *d e*

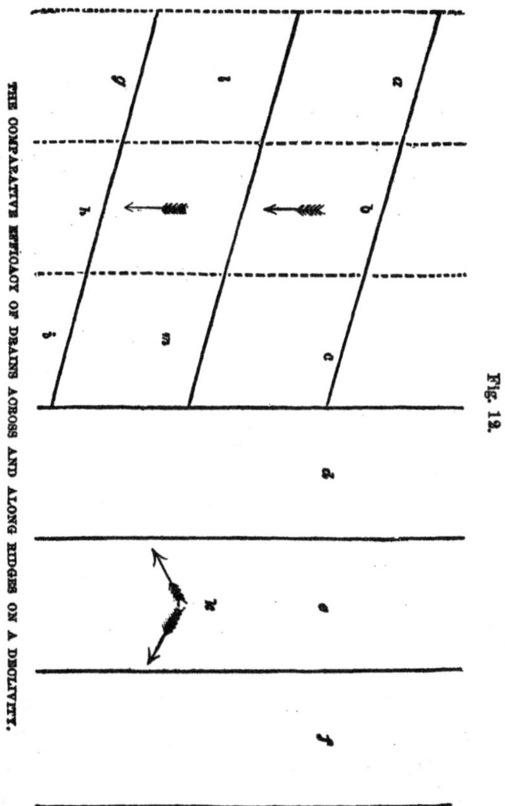

Fig. 12.

THE COMPARATIVE EFFICACY OF DRAINS ACROSS AND ALONG RIDGES ON A DECLIVITY.

f, having the drains parallel to them, and up and down the inclination of the ground, the water will take a diagonal direction towards the bottom of the drains, as indicated by the deflected arrows at *k*. Whereas on the ridges *a b c*, which have oblique drains, *a*, *l*, *g*, Fig. 12, the water will have to run in the direction of the arrows *b* and *h*, in doing which it will have to traverse the entire breadth of the ground betwixt *a* and *l* or *l* and *g*, just double the distance the other drains have. Mr. Thomson, of Hangingside, Linlithgowshire, drained 150 acres of land having an inclination varying from 1 in 10 to 1 in 30. Portions of three fields had drains put into them in 1828, 1829, and 1830, in the oblique direction, and, finding them unsuccessful, he put them in the direction of the slope, like the rest of the fields. 'In order,' says he, 'to ascertain the cause of these failures, a cut was made in the field first referred to, entering at a given point, and carrying forward a level to a considerable depth, when it was clearly seen that the substrata, instead of taking in any degree the inclination of the surface, lay horizontally, as represented in Fig. 5.' "*

It will be observed, from what has been said relative to the direction of drains with regard to the inclination of the surface, that *if the land slopes in different directions, then each plane of inclination should have a system of drainage for itself.*

With regard to the *positions of the main drains*, as they are intended to carry away accumulations of water from the smaller, they should *occupy the lowest parts of the land.* If the field is so flat as to have very little fall, the water may be drawn toward the main drains by making them deeper than the other drains, and as deep as the fall of the outlet will allow. If the field have

* Stephens.

a uniform declivity one way, one main drain at the bottom will answer every purpose; but, should it have an undulating surface, every hollow of any extent, and every deep hollow of however limited extent, should be furnished with a main drain. No main drain should be put nearer than five yards to any tree or hedge that may possibly push its roots toward it.

The *small drains* should generally be placed nearly at right angles to the main drains, and in lines parallel to each other, down the declination of the ground; not that all the drains of the same field should be parallel to one another, but only those in the same plane, whatever number of different planes the field may consist of. In a field of one plane, there can be no difficulty in setting off the small drains, as they should all be parallel, and all terminate in the same main drain, whether the field is nearly level or has a descent.

Fig. 13.

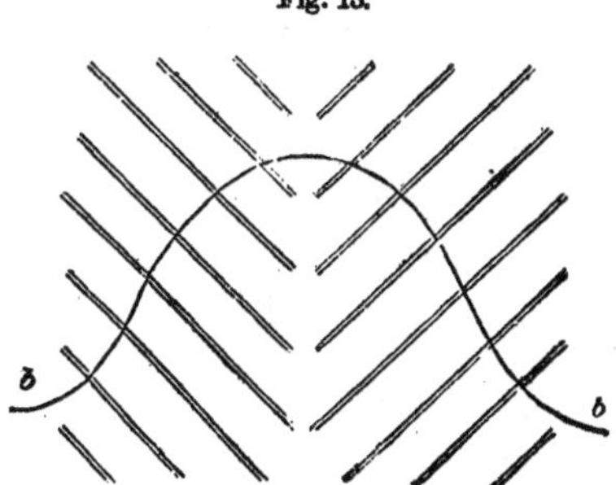

PARALLEL DRAINS IN ACCORDANCE WITH THE SLOPE OF THE GROUND.

It is not necessary, in all situations, to place the small drains precisely at right angles to the main drain. Where the surface varies considerably, it will be proper to let the position of the small drains be adapted in places to the fall of the surface. For instance, when the field has an undulating surface, a main drain is carried up the hollowest part of it, and the small drains are brought in parallels down the inclination to it.

Such drains should be cut, as in Fig. 13, up and down the inclined surface *b b*, toward the main drain, which would occupy the line along the points of junction of the drains *b b*.

The subjoined *sketch of a field thoroughly drained*, as to the *direction* and *position* of the drains, will elucidate the foregoing explanations: *a b* is the main drain formed in the lowest head-ridge; and if the field were of a uniform surface, the drains would run parallel to one another from the top to the bottom into the main drain, as those do from *a* to *c*, connected as they should be at the top with the drain *d e* running along the upper head-ridge. But as there may be inequalities in the ground, a very irregular surface cannot be drained in this manner, and must therefore be provided with sub-main drains, as *f g* and *h i*, which are each connected with a system of drains belonging to itself, and which may differ in character from each other, as *f g* with a large double set *k l* in connection with it, and *h i* with only a small single set *m*; the sub-main *f g* is supposed to run up the lowest part of a pretty deep hollow in the ground, and the drains *k* and *l* on either side of it are made to run down the faces of the declivities as nearly at right angles to the sub-main as the nature of the inclination of the ground will allow, so as always to preserve the natural tendency of water to find its way down the hollow. There is also a supposed fall of the ground from the height above *l* toward *k*, which causes the drain at *m* to run down and fall into what would be a common drain *h i*, were it not, from this circumstance, obliged to be converted into a sub-main. The sub-main *f g* may be

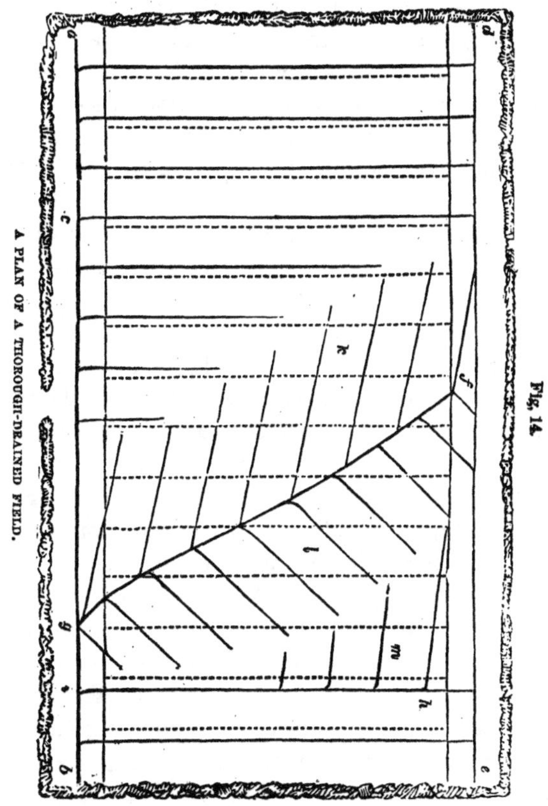

Fig. 14.

A PLAN OF A THOROUGH-DRAINED FIELD.

made as large as the main drain *a b*, as both have much to do; but the sub-main *h i* may be made comparatively smaller, and not larger, from the top of the field, than a common drain, until it reaches the point *h*, where the collateral drains begin to join it. The main drain should be made larger below *g* to *i* than above it, and still larger from *i* to *b*, which is its outlet. It will be observed that all the common drains *a* and *c*, and at *l* and *m*, have their ends curved, those at *k* not requiring that assistance, as they enter more obliquely into the main, from the position of the slope of the ground. The dotted lines represent the upper and lower head-ridges, and the open furrows of the ridges of the field; and it will be observed that the drains are not made to run in the open furrows—that is, the black lines in conjunction with the dotted—but along the furrow-brows of the ridges. This is done with the view of not confounding the open furrows and drains in the figure; but it is a plan which may be followed with propriety in subsoils otherwise than of strong clay; that is, of a light loam resting on a rather retentive subsoil; the water falling upon which should not be drained away by the small drains receiving it through their tops, but rather by the absorption of the water toward them from below the ploughed soil, as far as the subsoil is porous. A hollow, such as that occupied by the sub-main drain *f g*, also indicates that the soil is a loam, and not strong clay. Although the ridges are supposed to be fifteen feet wide, they bear no true relation to the size of the field; so that this diagram should not be considered as showing the relative proportions of the distances betwixt the drains and the size of the field.*

The *fall*, or *angle of declination* at which drains are constructed, requires to be guarded, especially in two particulars, either of which, although opposite in their

* Stephens.

effects, are equally prejudicial to their efficiency. 1st. In even-surfaced grounds there is danger of their being constructed too nearly on a level to allow the water to flow freely through them; 2d. In hilly land there is danger of the declivity of their position producing so rapid a current in wet seasons as to injure them by the friction of the water on their sides and bottom, and by their becoming in consequence choked up by the soil thereby loosened, which will often cause them to burst, or be *blown,* as it is technically called by drainers.

Whatever be the kind of substance over which water flows, it has the power of abrading it; for, besides earthy matter, it will in time wear down by friction the hardest rock. It seems, however, to be a prevalent opinion, that hard clay can for any length of time withstand the action of water in a drain. Were clay, indeed, always to retain the hardness it at first exhibits, it would require no protection from the abrading action of water; but, when it is known that it cannot possibly remain so, the safest practice is to afford it protection by a covering, such as a flat stone or tile, both of which obtain the name of *drain-soles.* The effects usually produced by water on clay subsoils are, that the lowest stratum of stones and the tiles become imbedded in it to a considerable depth, as has been found to be the case when drains that have *blown* have been reöpened. In somewhat softer subsoils, the sandy particles are carried along with the water, and deposited in heaps in the curves and joinings of drains; and, where the subsoil happens to be more sandy than clayey, the foundation which supports the bottom of the drain, whatever be the material it is made with, gives way, and the matter thus displaced forms obstructions at parts which render the drain above them almost useless. Water also carries sand down the sides of the drain, and, where there is no duct, deposits it among the lowest stratum of stones.*

Fig. 15

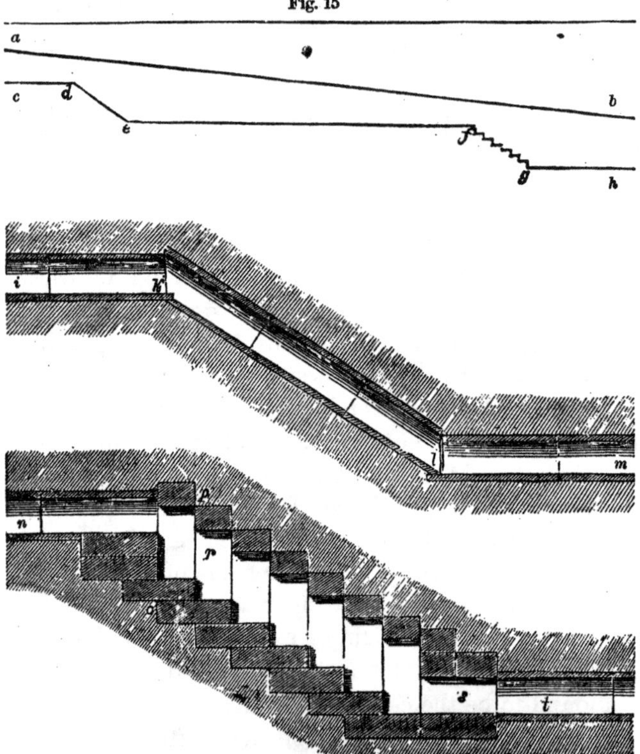

THE DIFFERENT FORMS OF CONDUITS IN THE INCLINED PLANES OF DRAINS.

Should it so happen, from the nature of the ground, that the *fall in the main drains is too rapid* for the safety of the materials which construct them, it is easy

* Stephens.

to cut such a length of the proper fall as the extent of the ground will admit—cutting length after length, and joining every two lengths by an inclined plane. The inclined planes could be furnished with ducts like the rest of the drain, or, what is better, in order to break the force of the water, like steps or stairs, of brick or stone masonry, built dry. Fig. 15 will illustrate this method at once, where *a b* represents the line of the lowest fall that can be obtained for a main drain in a field; but which is very considerable, and much more so than a main drain should have which has to convey a quantity of water. To lessen the fall, let the drain be cut in the form represented by the line *c h*, which consists of, first, a level part at the highest end, *c d*; then of an inclined plane, *d e*; again of a level part, *e f*; again of an inclined plane, *f g*; and, lastly, of a less level part, *g h*, to allow the water to flow rapidly away at the outlet; and this part may be parallel with the inclination of the ground.*

In the case of *level grounds*, as main-drains occupy the lowest parts, the *fall* in them cannot be so great as in other parts of the field. In such cases, the fall may entirely depend on cutting them deeper at the lowest end; but, when the fall is small, the duct should be larger than when it is considerable; because the same body of water will require a longer time to flow away. "People frequently complain," says Mr. Smith, "that they cannot find a sufficient fall, or *level*, as they sometimes term it, to carry off the water from their drains. There are few situations where a sufficient fall cannot be found, if due pains are exercised. It has been found in practice, that a water-course, thirty feet

* Stephens.

wide and six feet deep, giving a transverse sectional area of 180 square feet, will discharge 300 cubic yards of water per minute, and will flow at the rate of one mile per hour, with a fall of no more than *six inches per mile*." On the principle of the acceleration of water from drains, main drains, where practicable, should be six inches deeper than those which fall into them; and the greater depth has the additional advantage of keeping the drains clear of sand, mud, or other substances which might lodge, and not only impede, but dam back the water in the drains.

Should the fall from the mouth of the main drain to a river be too small, and there be risk, at times, of the overflowings of the river sending back-water into the drain, the drain should be carried down as far by the side of the river, as will secure a sufficient fall for the outlet. Rather be at the expense of carrying the drain *under* a mill-course or rivulet, than permit back-water to enter it.*

We now proceed to inquire the *depth* to which drains should be cut, and their *size*. These two branches of the subject depending, to some extent, the one on the other, will be best considered together. Mr. Stephens thus speaks of the size of drains: "A drain is not a mere ditch for conveying away water; were it only this, its size would be easily determined by calculation, or experiment, of the quantity of water it would have to convey in a given time. But the principal function of a drain is to *draw*† water toward it from every direc-

* Stephens.

† The term "*draw*," although in common use with drainers, is inaccurate, as it indicates a condition in a drain implying a force to attract water towards it. What is really meant by it, is the greater

tion; and its secondary purpose is to convey it away when collected; though both properties are required to be present, to insure the drain performing its entire functions. These being its functions, it is obvious that the greater the area its sides can present to the matter out of which it draws water, it should prove the more efficacious; and it is also obvious, that this efficiency is not so much dependent upon the breadth, as upon the depth of the drain; so that, other things being equal, the deeper a drain is, it should prove the more efficient. Now, what are the circumstances that necessarily regulate the depth of drains? In the first place, the culture of the ground affects it; for were land never ploughed, but in perpetual pasture, no more earth than would support the pasture grasses, would be required over a drain, and this need not, perhaps, exceed three inches in depth. The plough, however, requires more room; for the ordinary depth of a furrow-slice is seldom less than seven inches, and, in cross-furrowing, eight inches are reached, and two inches more than that, or ten inches in all, may suffice for ordinary ploughing; but, in some instances, land is ploughed with four horses instead of two, in which case the furrow will reach twelve inches in depth, so that fourteen inches of depth will be required to place the materials of the drain beyond the danger of an extraordinary furrow. But farther still, subsoil and trench ploughing are sometimes practiced, and these penetrate to sixteen inches below the surface,

facility presented by the porous material of the drain for water to run into it; and it is on that account that the drains are filled for some distance above the water-passage, or duct, with stones or porous material; but it will be seen, in the subsequent pages, that the capacity of drains to act *in this way*, is now much disputed.

so that eighteen inches of earth, at least, you thus see, will require to be left on the top of a drain, to place its materials beyond the dangers arising from ploughing. This depth having been thus determined by reference to practice, it should not be regarded as a source from which a supply of moisture is afforded to the drain by its drawing power, the water only passing through it by absorption; for it is certain that ploughed land will absorb moisture, whether there be any drain below it or not. The *drawing* portion of the drain must, therefore, lie entirely below eighteen inches from the surface. Now, it will be requisite to make the drain below this as deep as will afford a sufficient area for drawing powers of the lowest degree among subsoils. And what data do we possess to determine this critical point? In the first place, it is evident that a subsoil of porous materials will exhaust all its water in a shorter time than one of an opposite nature. Judging from observation, I should say, that one inch thick of porous materials, will discharge as much water, in a given time, as six inches of a tilly, or any number of inches of a truly tenacious subsoil. What conclusions, then, ought we to draw from these data? Certainly these: that no depth, beyond the upper eighteen inches farther than what is required for the materials of the drain, will draw water from a truly tenacious subsoil, and that it is, therefore, unnecessary to go any deeper in such a subsoil; that it is also unnecessary to go any deeper in a subsoil of porous materials, because a small depth in it will draw freely; and that it is only requisite to go deeper in the intermediate kinds of subsoil. Still, you have to inquire what should be the specific depths in each of these cases? In the case of really tenacious subsoil, the size of the duct for the water depends on the quantity to pass through it; but, giving the largest allowance of six inches, with a sole beneath and covering above, one foot seems ample depth for these materials to occupy, so that a drain of two and a half feet seems sufficient for the circumstances attending such a subsoil, that is, its minimum depth, which, in such a case, may also be held to be a maximum. In the case of a porous subsoil, it is absolutely necessary for the preservation of its loose materials in their proper position, to have a lining of artificial materials as far as these extend; and as such a lining can hardly be constructed of sufficient strength of less depth than one foot, it follows that two and a half feet is the minimum depth also in such a subsoil; but there is this difference betwixt this subsoil and the tenacious one, that the porous may be made as deep as you please, provided you apply sufficient materials for the support of the loose materials. With regard to tilly subsoils, since one foot is requisite for the safety of the filling materials, it does not seem an overstretch of liberality to give six inches more for extension of the drawing surface, so that the minimum depth, in this case, seems to be three feet, and as much more as the peculiar state of the subsoil, in regard to tenacity and porosity, will warrant you to go."

"The next step is to fix the *depth of drain most suitable for draining the particular field;* and this can only be done by having a thorough knowledge of the nature of its subsoil. I have already given reasons for fixing the minimum depth of drains in the different kinds of subsoil; but, as the reasoning given there only establishes the principle, it is not sufficient to determine the most proper depth for every peculiarity of circumstances; for this must be determined by the nature of the subsoil which guides the whole affair. If the field present an uniform surface, but inclining, let at least two exploratory drains be cut from the bottom to the top of the field, if its extent does not exceed ten acres, and as many more as it is proportionally larger; and if the subsoil of both is found at once tilly, that is, drawing a little water, let the cut be made three feet deep without hesitation. On proceeding up the rising ground, the depth may be increased to four feet, to ascertain if that depth will not draw a *great deal more water* than the other. Should the subsoil prove of porous materials, two and a half feet—the minimum—may suffice; though, on going up the rising ground, it may be increased to three feet, to see the effect; but should it, on the other hand, prove a pure tenacious clay, two feet will suffice at first, increasing the depth in the rising ground to two and a half and even three feet; for it may turn out that the stratum under the tenacious clay is porous. Where the surface is in small undulations, the drain should be cut right through both the flat and rising parts. In very flat ground, any considerable variation of depth is impracticable, and only allowable to preserve the fall. From such experimental drains data should be obtained to fix the proper dimensions of the other drains."

"If you find the substratum pretty much alike in all the experimental drains, you may reasonably conclude that the subsoil of the whole field is nearly alike, and that all the drains should be of the same depth; but, should the subsoil prove of different natures in different parts, then the drain should be made of the depth best suited to the nature of the subsoil. A correct judgment, however, of the true nature of the subsoil, cannot be formed immediately on opening a cut; time must be given to the water in the adjoining ridges to find its way to the drain, which, when it has reached, will satisfactorily show the place which supplies the most water; and, if one set of men open all the cuts, by the time the last one has been finished, the first will probably have exhibited its powers of drawing; for it is a fact that drains do not exhibit their powers until some hours after they have been opened. When you are satisfied that the drains have drawn in dry weather as much water as they can, you will be able to see whether or not the shallowest parts have drawn as much as the deepest; and you should then determine on cutting the remainder to the depth which has operated most effectually. If rainy weather ensue during the experiment, still you can observe the comparative effects of the drains, and abide by the results. Never mind though parts of the sides of the cuts fall down during dry or wet weather; they need not be regretted, as they afford excellent indications of the nature of the subsoil, the true structure of which being left by the fall in a much better state for examination than where cut by the spade; and you may then observe whether most water is coming out of the highest or lowest part of the subsoil. It is essential for the durability of drains to bear in mind that they should always stand, if practicable, upon impervious matter, to prevent the escape of the water from the drain by any other channel than the duct."*

* Stephens.

With reference to the *size of drains*, upon the *thorough drainage* principle, it is considered that for the small drains, tiles of two inches diameter in the bore are large enough to carry off any quantity of water that is practically found to be present after the wetest seasons. And with regard to main or *other drains for cutting off springs*, &c., their size must be regulated by the quantity of water to be discharged. It has been proved in practice that a water course thirty feet wide and six feet deep will discharge 300 cubic yards of water per minute, and will flow at the rate of one mile per hour with a fall of only six inches per mile. For thorough main drains, Smith says, with a fall of not less than one foot in 100 yards, a drain ten inches wide and twelve inches deep, will void the rain water from 100 acres.

The *distance apart* at which drains should be constructed remains for us to consider. It will be evident that the distance of drains from each other, involves materially the question of the expense of making them; and that the object therefore is to place them at as great a distance apart as they can be placed, to be truly effective. Upon this point Mr. Stephens says:

"The *distance* that should be left between the drains can only be satisfactorily determined after the depths of the drains have been fixed upon, as drains in a porous substratum, which draw water from a long distance, need not, of course, be placed so close together as where the substratum yields water in small quantities; and as drains may be of different depths in the same field, according to the draining powers of the substratum, so they should be placed at different distances in the same field.

"In a partially impervious subsoil, fifteen feet are as great a distance as a 3-feet drain can be expected to draw; and, in some cases, a 4-feet one will be required. In more porous matter, a 3-feet drain will probably draw twenty feet, with as great effect; and in the case of a mouldy, deep soil, resting on an impervious subsoil—a drain passing through the mould, and resting four inches in the impervious clay—which may altogether make it four feet deep—will draw, probably, a distance of thirty feet."

Mr. Stephens further says, in speaking of drains for stiff clay:

"I know it is a common impression among farmers, that if a subsoil cannot draw water, there is no use of making drains *in* it, and this opinion I conceive to be quite correct in regard to pure clay subsoils, which cannot draw water at all. But the view I take of the matter is this, that pure clay subsoils are very limited in extent, and that many clays which *seem* quite impervious may draw water notwithstanding. Admitting that the subsoil draws water at all, which is the supposition in the present case, it is clear that the larger the area is extended for drawing it, the more water will be drawn into the drain. Now, a large area can only be secured by making drains deep and close together; and in the case supposed above, it appears to me that three feet in depth, with fifteen feet asunder, will not give a greater area than is requisite for drawing water out of such ground. When, on the other hand, the subsoil is free, and discharges water as freely, so large an area is not required to dry the subsoil, and drains of less depth and at greater distance will answer the same purpose as in the other case, such as thirty inches in depth and thirty feet asunder. You must endeavor to make the depths and distances of the small drains suit the nature of the subsoil, for it is impossible for me to lay down here any absolute rule in a matter which admits of such diversity of character."

The disadvantage of a mistaken economy in the construction of thorough or frequent drains too far apart, has been thus pointed out:

Conceiving that a drain in every furrow, in a tilly subsoil, is attended with more expense than any increase of produce would warrant, a farmer in East

Fig. 16.

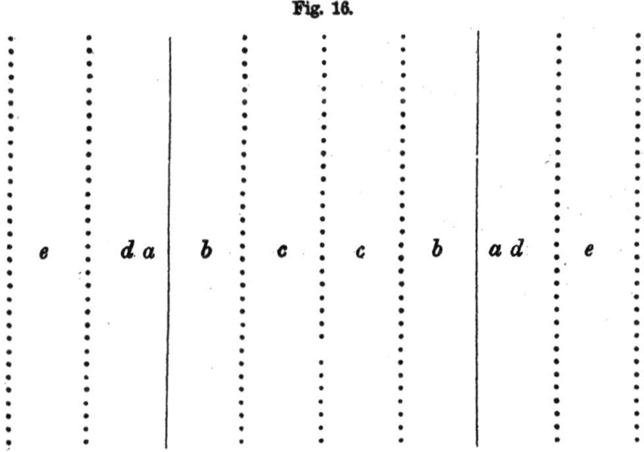

THE EFFECTS OF TOO GREAT A DISTANCE BETWIXT DRAINS.

Lothian put a drain in every *fourth* furrow; and he caused them to be cut four feet deep. A figure will best illustrate the results where the black lines *a*, Fig. 16, are the drains between every fourth furrow, and the dotted lines represent the intermediate undrained furrows; and where it is evident, at the first glance.

that the drains *a* have to dry two ridges on each side *b c* and *d e*, of which we should expect that the two ridges *b* and *d*, being nearest to *a*, should be more dried, in the same time, than the two farthest ridges *c* and *e*, and the result agrees with expectation; but still, had the subsoil been of an entirely porous nature, both ridges might have been sufficiently dried by *a*. But mark the results of this particular experiment. The two ridges *b* and *d*, nearest to *a*, actually produced nine bushels of corn more per acre than the two more distant ridges *c* and *e*. This is a great difference of produce from adjoining grounds under the same treatment, and yet it does not show the entire advantage that may be obtained by drained over undrained land, because it is possible that the drain *a* also partially drained the distant ridges *c* and *e*; and this being possible, together with the circumstance that none of the ridges had a drain on each side, it cannot be maintained that either the absolute or the comparative drying power of these four-feet drains was exactly ascertained by this experiment.* It may be conceived, however, that if the drains had been put into every other, instead of every fourth furrow, that the produce of all the ridges would have been alike, inasmuch as every ridge would then have been placed in the same relative position to a drain.

The above lengthened quotations from Mr. Stephens' able paper on Drainage, have been given because they express well the opinions of the advocates of the system of frequent drainage, recommended by Smith of Deanston; and together with the preceding observa-

* Quarterly Journal of Agriculture, vol, viii.

tions on the situation of drains, and their depth and distance, they set forth the principles on which the system is founded. For many years after its introduction by Smith, his system to a great extent superseded the use of Elkington's deep drainage, and by many eminent men this is still held to be the perfection of drainage for all land with stiff subsoils. There has sprung up, however, within the last few years, a class of draining engineers, of equal eminence and experience, who are very strenuous in their recommendation of a modification of the above plan. This subject has been before referred to, and this seems the fitting place, as there appears much truth in what these gentlemen advance, to introduce their opinion and advice to the reader.

The gentlemen alluded to, found their recommendations upon an opinion that the supposition that the sides of drains "draw" water into them is erroneous; (*see note, page* 60,) and that all the water from the surface sinks perpendicularly, (or as nearly so as the heterogeneous nature of the substrata may permit,) and enters the drain at its bottom only. They do not differ materially from Smith's system as to the *position* of the drains which they consider should go up and down the slope, but they advise that in deep soils four feet should be the *minimum* depth, and that they may be placed at from twenty to fifty feet, or in some soils much further apart; the stiffness of the soil in some situations requiring them nearer together than in others.

Mr. Josiah Parkes originated this departure (which in some respects leans to the Elkington deep drainage, but is still distinct from it,) from Smith's plan, and at present it is upheld by Mr. J. Bailey Denton and Mr.

Hewitt Davis, two of the most successful draining engineers of the day, and by very many others—both scientific and practical men; and although Smith's old plan is by no means in want of experienced and intelligent supporters amongst the same class of men, the weight of evidence appears to uphold the correctness of the opinion that Parkes has put forward. Another feature that Mr. J. B. Denton (who is undoubtedly one of the most talented of the disputants) advises is, that always keeping the course of the small drains with the slope of the surface, they should not be confined to that perfect uniformity of relative position which is adopted by Smith, (and which Mr. Denton has termed the "Gridiron plan,") but that where the character of the soil alters so as to be more open and porous in one part of a field than in another, and where the surface level varies materially, that the position and relative distance of the drains should be varied also, so that cutting useless drains may be avoided, but the direction with the slope still retained.

These remarks will show that the cutting and mode of constructing the drainage is the same, whichever opinion of these two classes of drainers be acted upon; and that it is only to be looked at upon the question of principle as to comparative efficiency.

The plan of frequent drainage as advanced by Mr. Smith, was first put forward in consequence of the indiscriminate adoption of Elkington's deep drainage in all situations, being found to fail in many instances; although it is well supposed that in some of those cases the fault was rather in the want of judgment in its application than in the system itself. Adhering to the beneficial parts of Mr. Smith's system, Parkes and his followers of the present day differ from him as to the way in which the drainage is effected; that is, whether the drains "draw" towards their sides the water in its passage downwards from the surface, or whether the bulk of the water passes down to a given depth (regulated by the depth of the drains as deeper drainers contend), and so enters the drains at their base. And acting upon the belief (which seems the correct one) that it does so, they consider that there is a saving of expense, and also a more beneficial result to be attained, by a deeper position of the drains, than by that advanced by Mr. Smith.

This result they do not attribute simply to the efficiency of the drainage, but also to the principle of converting the water falling on the surface into an agent for fertilizing the ground, by its passage through a greater depth, and in its passage imparting to the particles of soil certain elements of nutrition for vegetation which it is known to contain.

A review of the opinions of the most eminent draining engineers of the present day, and a comparison of the various systems practiced, with the history of their gradual introduction and operation, undoubtedly leads to the conclusion that the most perfect manner of draining now practiced is by Elkington's system for the cutting and removal of springs, and subterraneous accumulations of water by boring, in some situations combined with the frequent or thorough draining system upon the deep principle advocated by Mr. Bailey Denton; and in other places, that which adopts the latter system alone, modified as regards the distance of the drains apart by the nature of the soil, and the conformation of the surface to be drained.

CHAPTER V.

SURFACE DRAINAGE.

SURFACE DRAINAGE consists in cutting channels or ditches, which are left open at the top for the water to run off. As compared with the systems before described, this one is very deficient.

Open drains, unless cut sufficiently deep to act on the principle of sub-surface drains, cannot so completely relieve the soil from the surplus water contained in it. They require, also, to be constantly cleaned out and repaired, as, from their exposure to all the vicissitudes of the weather, they are liable to fill up more or less. And they occasion a great loss of land, from the large aggregate area of their open surface. There are some situations where, on account of their less cost, they may be used, and others in which, in recently reclaimed lands, or on the score of difficulty in obtaining labor for more efficient works, the plan may nevertheless be adopted temporarily, if not permanently.

This mode of draining does not profess to interfere with any water that exists under the surface of the ground, farther than what percolates through the ploughed furrow-slices, and makes its way into the open furrows of the ridges. For the purpose of facilitating the descent of water into the open furrows, the ridges are kept in a bold, rounded form; and that the open furrows may be suitable channels for water, they are cleared out with the plough after the land has been otherwise finished off with a crop. The small channels cut with the spade are made through every natural hollow of the ground, however slight, and the water-furrows cleared into them at the points of intersection. The cuts are continued along the lowest head-ridge furrow, and cut across the hollowest parts of the head-ridge into the adjacent open ditch. The récipient ditch forms an important part of this system of draining, by conveying away the collected waters, and is made four or five feet in depth, with a proportional width. It is immediately connected with a larger open ditch, which discharges the accumulated waters from a number of recipient ditches into the river or lake, or other receptacle which is taken advantage of for the purpose. The large ditch is from six to ten feet in depth, with a proportional width. It is evident this is a system only applicable to soils that retain water for a long time on the surface.

The principles before described for sub-surface drainage, as to the situation and direction, apply nearly equally well to open drains. The damage to open drains from friction by the water, and the sand, and other matters held in suspension by it, is much greater in open than in covered drains. It will be found advantageous, in cases where large quantities of water are likely to pass through open drains, to cut the bottom of them in a semi-circular form, and wide in proportion to their depth; because, the quantity of water passing through being the same, the abrasion is less on the bottom of a wide drain than in a narrow one, owing to the shorter height of the superincumbent column of water on any given part of the bottom. In some parts of the south of England, where, for particular local reasons, open drainage is used, open ditches, six or eight feet wide, are employed, instead of fences, which in some measure lessens the loss of the land occupied by them.

In the adoption of this description of drainage, as its operation is always open to observation, the first cutting of them need not be on an extensive scale, because it is, of course, easy to add to the number at any time, if those formed are found inadequate to the purpose intended.

One of the most useful ways in which open drainage can be made available, is in the reclaiming of salt marsh lands, in which, from the great extent generally of the surface at which the wetness enters upon the land, it is desirable that the drainage should remain open until it be ascertained by experience that it is sufficient for the intended purpose. Afterwards, if wished, it may have ducts built, and be converted into under-drainage.

PART II.

PRACTICAL INSTRUCTIONS FOR THE CONSTRUCTION OF DRAINAGE.

THE examination of the land having been made, and the kind of drains being determined on, the practical operations of the drainer commence.

The first thing to be done in the formation of drains, is to ascertain the relative *levels* of the various parts of the land, so as to be certain that, when constructed, they will act efficiently in permitting the passage of the water along them. In some situations, where the inclination of the surface is considerable, no doubt can exist as to that. But where the surface is nearly level to the eye, or where it varies much in shape, the question can only be determined by taking accurate levels of the different parts, and thereby ascertaining the various relative heights, with regard to the outlet by which the drainage water is to be carried off.

The next point is to determine the *materials* of which the drain shall be constructed. This must depend frequently upon the situation and the comparative value of the several available materials, and also upon the kind of drain to be made. These questions disposed of, the cutting of the drains may be commenced; and if underground drains, the culverts or ducts built or formed, and the drains again filled up to the surface, which completes the work.

These several matters will now be considered in detail.

CHAPTER VI.

LEVELLING.

LEVELLING is a branch of surveying, the object of which is to find a line parallel with the horizon, from which the rise and fall of ground may be measured. Its use to the drainer is to ascertain the difference of the heights of ground, so as to find the best descent for the drains; and, in level districts of country, it is often requisite, in order to know, before commencing drainage, whether one or another place will present the best fall for carrying off the water.

For these purposes, a simple mode of levelling is sufficiently accurate, except for the general drainage of extensive districts, when competent engineers would be employed; and it is only necessary here, therefore, to state the way in which what is requisite can be effected without the aid of expensive instruments.

Fig. 17.

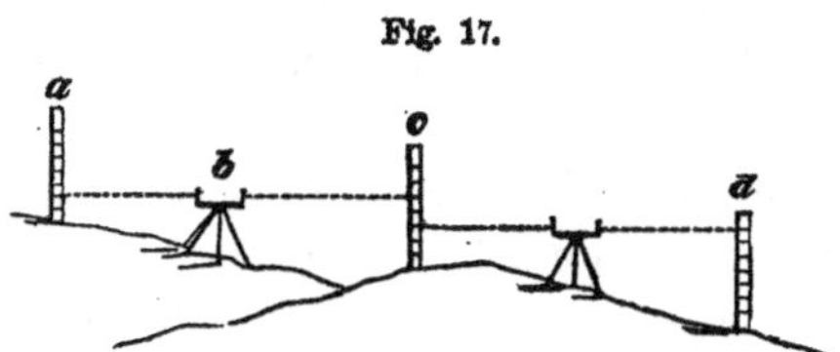

If a level be placed at *b*, Fig. 17, in a perfectly horizontal position, and two staves, marked on one side with feet and inches, be placed at *a* and *c*, a person, looking along the level from *b*, first to *a*, and noting the number of feet at which a straight line, drawn from *b*, would cut the staff at *a*, and then looking towards *c*, and noting where the same line would cut *c*, will have only to deduct the one number of feet from the other, and the difference will be the height of the ground at *a* above the ground at *c*. If then the level be moved on between *c* and *d*, and the same operation be repeated, the same result will, of course, be obtained between these two points, and so on as often as need be. Whatever may be the distance, by adding all the observations taken in *each* direction together, those seen looking backwards in one column, and those seen looking forwards in another, and subtracting the sum of one from the other, the *difference* will give the elevation of the one of the two extreme points of observation above the other. If the two columns add up precisely alike in amount, there is no difference in the level of the starting point from that at which the observations terminate. By this means, it will be known with certainty whether a fall can be obtained for carrying off drainage water at any given place from ground near or distant from it.

Care must be taken, in moving the level from place to place, for making the observations, that the sight is always the same distance from the surface of the ground on which it stands. For great distances, the sights must be taken, and the marks on the staves read off by the aid of a telescope, attached to the level; but for drainage of land for agricultural purposes, the observations can be made at such distances only as the eye can reach without the assistance of that instrument. That being the case, the levels may be taken by means of a simple water-level, which has been long in use, and of which the following description is taken from Mr. J. J. Thomas's small volume,* which, although of very unpretending character, contains information which would probably save many a farmer ten times its cost every year of his life, by teaching him the *principles* on which he *would use, to the best advantage*, his labor-saving machinery and implements:

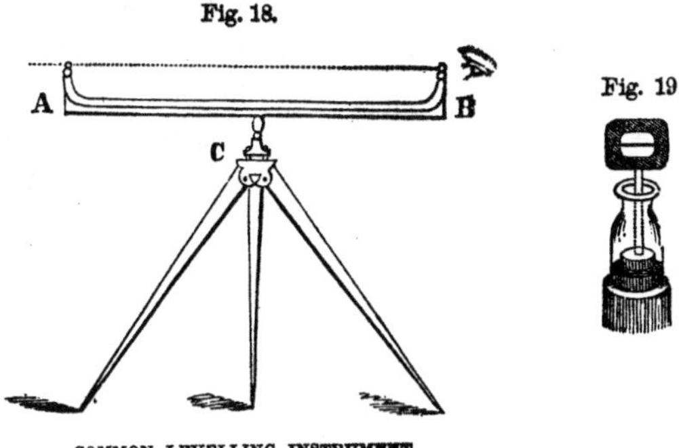

COMMON LEVELLING INSTRUMENT.

" The water level is made of a lead tube, about three feet long, bent up an inch or two at each end, and stiffened by fastening to a wooden bar, A, B (Fig. 18). Into each end is cemented, with sealing-wax, a small and thin phial, with the bottom broken off, so that when the tube is filled with water it may rise freely into the phials. If the tube be now filled with water, colored with cochineal, or any dye-stuff, and then placed upon the tripod, C, by looking across the two surfaces of liquid in the phials, an accurate level may be obtained.

* " Farm Implements, and the principles of their construction and use." By John J. Thomas.

When not in use, a cork is placed into each phial. "Sights" of equal height, fastened to pieces of cork floating on the water, as shown in Fig. 19, give a more distinct line for the eye. The sights are formed of fine threads, or hairs, stretched across the square openings. To ascertain whether these threads are both of equal heights above the water, let a mark be made where they intersect some distant object; then reverse the instrument, or turn it end for end, and observe whether the threads cross the same mark. If they do, the instrument is correct; but if they do not, then one of the sights must be raised or lowered until it becomes so."

For the purpose of keeping a uniform grade of fall in cutting drains, a mason's level will answer; but no level is so useful for that purpose as the A, or span level, Fig. 20.

Fig. 20.

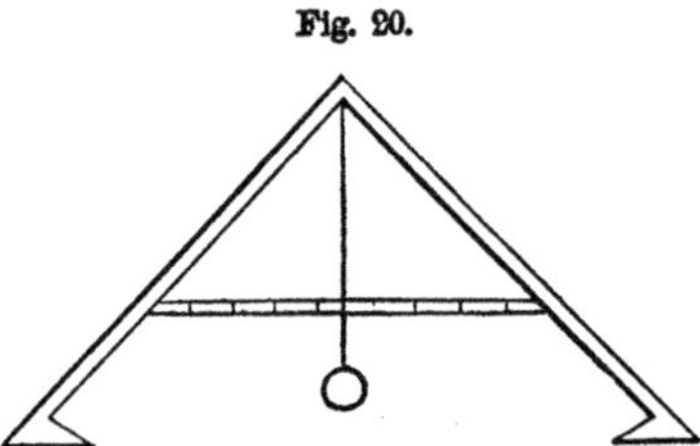

Such a level may be easily constructed of wooden laths. The span should be either sixteen feet six inches, or half that length. The two feet being placed on a perfectly level floor, the plumb-line will hang in the centre, where a notch should be made in the cross-bar. Then place a block of wood exactly an inch thick under one leg, and mark the place on the cross-bar that the plumb line then touches. Put a second block of one inch under the same leg, and again mark the place of contact of the plumb-line on the cross-bar, and so on, as far as requisite. Afterwards, put the blocks, one by one, under the other leg, and mark the bar on the other side of the centre. When thus prepared, if the span of the level be sixteen feet six inches, the plumb-line will indicate upon the bar, by the dis tance that it hangs from the centre notch, the number of inches *per rod* of the ascent or descent of any drain in which it is placed. If the span of the level be eight feet three inches, it will, of course, in the same way, indicate the number of inches of ascent or descent in half a rod.

By the aid of the above simple instruments, the drainer will be able to take the necessary levels for all ordinary works.

CHAPTER VII.

DIFFERENT KINDS OF DRAINS DESCRIBED.

A GREAT variety of drains are described in this chapter, some of which are by no means desirable, where more perfect ones can be adopted; but the question of expense must always be of importance in drainage, and that must depend upon the kind of materials to be had at reasonable cost. This may render it needful sometimes, whatever may be the system of drainage adopted, to select a less perfect description of drain than, but for its cheap cost, would be proper. From the many kinds of drains here mentioned, that selection can be made; but it is needless to remark, that the more durable the first construction, the more lasting and satisfactory will the work prove; and the subsequent expense in keeping it in order will be, likewise, proportionably less.

For main drains, open ducts, either stone, slate, or tile, should always be provided, if possible. *Stone drains* may be made with two flat stones, placed against each other at the bottom of the drain, with another covering, both, as at *a*, Fig. 21, forming an equilateral duct of six inches in the side. It should be held down in its position by small stones, *b*, to a height of eighteen inches; then covered with turf or other dry substance, *c*, and the earth *d* returned above them. In making this form of duct, of three feet in depth, the drain will require to be eighteen inches wide at top, to allow the drainer room to work while standing on the narrow triangular space at the bottom. Placing the apex of the triangle undermost gives the water power to sweep away any sediment along the narrow bottom; but it possesses the disadvantage of permitting the water to descend by its own gravity, between the joining of the stones, to the subsoil, which runs the risk of being softened into a pulp, or of its sandy portion being carried away; and it is possible for a stone to get jammed in the narrow gutter and form a damming.

Another form of duct, which is sometimes called a *till-drain*, may be seen in Fig. 22, where *a* is the duct,

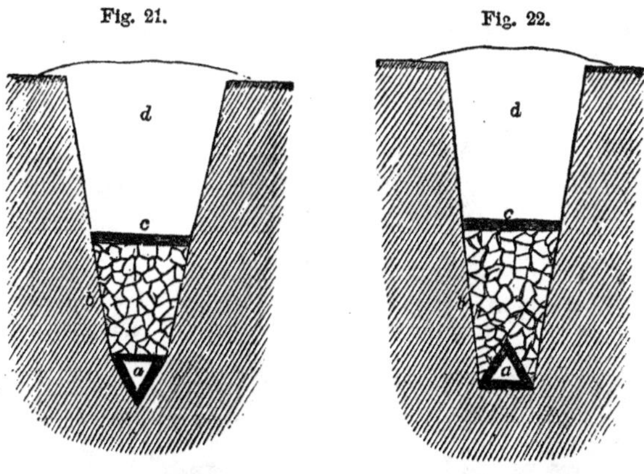

THE TRIANGULAR STONE DUCT. THE COUPLED STONE DUCT.

consisting of a sole lying on the ground, supporting two stones meeting at the top, forming an equilateral triangle of six inches a side. This form encourages a deposition of sediment to a greater degree than the former, but it prevents to any dangerous extent, the descent of the water under the sole. Having a flat bottom, the drain can easily be cast out with a width at top of only fifteen inches to a depth of three feet.

A *more perfect duct* than either of these is made by *a tile and sole.* In all main drains, formed of whatever materials, capable of conveying a considerable body of water, a sole is absolutely requisite to protect the ground from being washed away by the water, and a more effectual protection cannot be given to it than by tile and sole. A *main-tile*, of four inches wide and five inches high, will contain a large body of water; but should one such tile be considered insufficient, two may be placed side by side, as represented by *a* and *b* in Fig. 23. Should a still larger space be

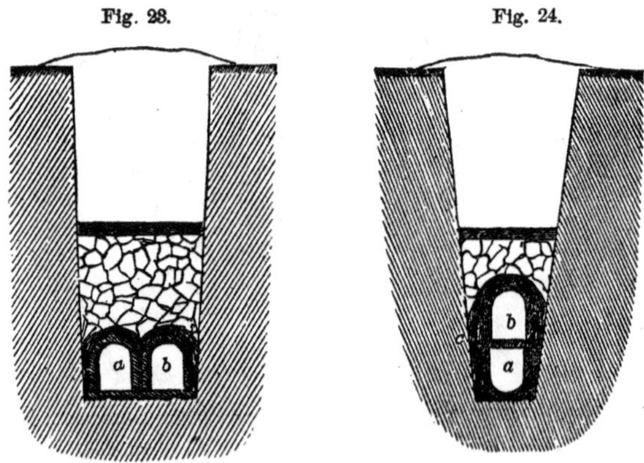

THE DOUBLE TILED MAIN DRAIN. THE INVERTED DOUBLE TILED MAIN DRAIN.

required, one or two soles may be placed above these tiles, and other tiles set on them, as *a* and *b* are. Or should a still deeper and heavier body of water be required to pass through a main drain, one or two tiles can be inverted on the ground on their circular top, as *a*, Fig. 24, bearing each a sole, *c*, upon its open side, and this again surmounted by another tile, *b*, in its proper position. In such an arrangement, there is some difficulty in making the undermost tile, *a*, steady on its top; for which purpose the earth is taken out of a rounded form, and the tile carefully laid and wedged round with stones or earth; but there is greater difficulty in making the uppermost tile, *b*, stand in that position without a sole, as is recommended by some writers on draining, because the least displacement of either tile will cause the upper one to slip off the edge of the under, and fall into it. In the narrowest of these cases of main drains with tiles, the drains can be easily cut at fifteen inches wide at top to the depth of three feet. Small stones should be put above the tiles, if at all procurable, to the height of eighteen inches above the bottom; if not procurable, gravel will answer the same purpose; and, if both are beyond reach, they should be enveloped with thin, tough turf.*

The following are various kinds of drains *adapted for small drains* and for the *sub-main;* and others for general drainage of the land.

The Figure 58† represents a *small drain filled with small stones.* This drain is thirty-six inches deep, nine inches wide at bottom, twelve inches at the top of the stones, and the stones eighteen inches deep.

The *small Tile-Drain* is represented in Fig. 53.‡ The drain-tile should rest on a sole placed at bottom of the drain, unless the tile be made with the sole attached.

*** Stephens.** **† See page 152.** **‡ See page 140.**

The Figure 25 represents a *tile and stone drain*, which is declared by every writer on and practitioner of draining to be the *ne plus ultra* of the art, though very few have adopted it, because, in cases where stones have to be quarried and broken, it is an expensive mode. A tile, *a*, rests on a sole; small stones are packed around the tile by the hand until they cover it, as at *b;* the remaining small stones, *c*, are put in by the drain-screen; a covering is either put above them, or small stones beaten down with the beater, and the earth returned upon them. The width of the bottom is seven inches, width of the top twelve inches, depth two and a half feet, composed of eighteen inches of earth, and twelve inches to the top of the covering of the stones.

Fig. 25.

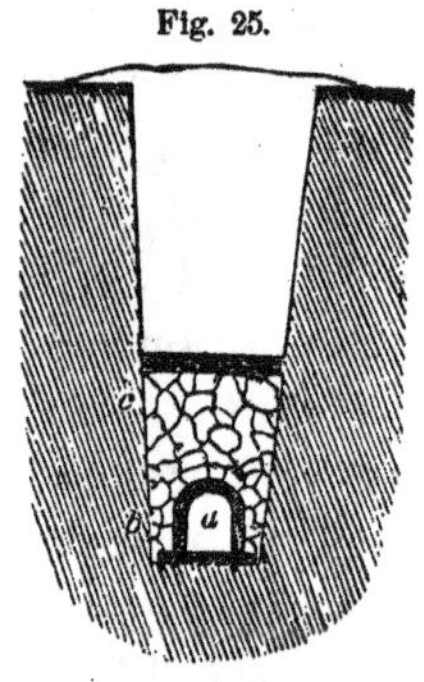

THE TILE AND STONE DRAIN.

A *drain* represented in Fig. 42* is *applicable to clay* land. The figure represents the drain as cut in the stiff soil with a flat stone, *d*, covering the duct, but before it is filled in with earth.

Mr. John S. Skinner, in some sensible remarks on draining, in his edition of the "Book of the Farm," has called attention to *a mode of draining with wood* which may be well deserving attention in places where stones cannot be procured. This plan (although not new) has been, Mr. Skinner says, well carried out by a Mr. Summers, near Nottingham, Maryland, who has reclaimed spots of land which had before been worthless, and which now yield heavy crops. He says: "The wood used for this purpose consists of the thinnings of

* See page 119.

plantations, *i.e.*, the small trees commonly converted into paling. Larch is preferable, on account of its greater durability; but Scotch fir, being the cheapest and most abundant kind in this quarter, is generally used. The drains to be filled with wood are usually thirty-two inches in depth, eighteen inches wide at the top, and about six inches at the bottom. It is essential to the efficiency and durability of wooden drains, that the sides be formed with a proper and regular slope from top to bottom. The small trees—or "spars," as they are designated—are prepared for being put into the drain in the following manner: A portion of the butt or thick end of each is sawn off for placing transversely in the drain, about six inches above the bottom; the breadth of the drain at this part may be assumed at nine inches, in which case the length of the cross-bars will require to be about fifteen inches, so as to have three inches resting on each side. They are generally about four inches in diameter, and are placed in the drains at intervals of four feet apart; they are forced firmly into their proper position by a few blows of a heavy mallet, the workman taking care that they are all in the same plane or level. Any earth loosened from the sides in striking down the bars is, of course, thrown out as the work is proceeded with. After the butt-ends of the trees (which are divested of their branches in the wood) are severed, and placed transversely in the drains in the manner just described, the remainder of them are laid longitudinally above the bars, three being commonly placed side by side, and covered with the branches and twigs, or with turf, heath, &c., previous to putting in the earth cast out in opening the drains. It is obvious that this method of draining can be adopted with advantage only in situations where timber is convenient and cheap, and when the subsoil is sufficiently cohesive to afford a proper support to the transverse bars of wood; hence it is inadmissible in the case of boggy lands. The putting in of the wood is accomplished in a very expeditious manner: two persons saw off the butts, and another places them in their proper position in the drain, after which the longitudinal spars are laid on as closely as possible, with the top and butt-ends alternately in the same direction, so as to make them fit the better. There is thus formed beneath the wood a channel for the passage of water, of about six inches in width and the same in depth. The cost of this mode of draining obviously depends much on the price of the wood employed. In most parts of this country the spars used for the purpose are obtainable at from 1s. to 1s. 6d. per dozen; and it requires four dozen, averaging twenty feet in length, to do a hundred yards of drain. Drains thus constructed have been known to last for a very long period; on one farm the writer has been assured that drains formed of wood in the manner just described have been in perfect operation for more than thirty years."

Brush Wood or *Bush Drains,* as they are called, are formed by laying down branches, spray, and brush wood in the bottom of the cuttings, to form the duct instead of tile or stone. Sometimes the brush is first laid down along the edge of the drain and formed into an endless cable a foot thick by binding it tightly together with willow, and afterwards it is rolled into the bottom of the cutting and trod down. This forms a more lasting drain than any other of this description. Usually the branches and brush are put into the cutting in a slanting direction with their root ends towards the bottom of the drain, and those ends also placed *with* the descent of the ground. They should be trodden down and compressed to half their bulk, and all brush drains should be covered with turf inverted on them before filling in. *The filling* should be executed on the same principles as for other drains.

A Slab Drain has been sometimes constructed, amongst the many modes in which wood has been used for drains; and an instance of this kind of drain is described in that excellent Farmer's newspaper the "*Country Gentleman*," which has been found to work well for several years. This drain is there described by Mr. J. Wilbur, of Beamis Heights, N. Y., as follows: "Our land was a retentive subsoil, and on such soil only would I recommend this kind of drain. We dug our drain of sufficient width at bottom to admit a common round shovel, and from twenty to thirty inches deep with moderately sloping sides. Then commencing at the upper end, we laid a common hemlock, bass wood, or other slab from ten to twenty inches in width, with the sawed side downward, and the upper edge reclining against the side of the ditch, so as to form a triangular throat between the side and bottom of the drain. We covered the irregular portions of the slabs with other pieces, and chinked with turf. We placed the slabs end to end, the same as tile are laid, and were careful to keep the throat clear as we advanced. We formed openings from the surface, wherever desirable, with open drains or dead furrows leading to them, all of which continued to work well at the time I visited the farm in 1852." (They were made twelve years previously).*

* Country Gentleman, vol. iii., page 311.

A Wooden Drain as a *substitute for Tile* has been proposed by a Scotch gentleman, which is represented in Fig. 26. He suggests larch as a wood which in wet

Fig. 26.

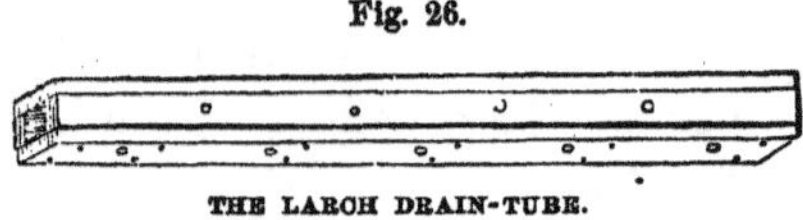

THE LARCH DRAIN-TUBE.

boggy soils is durable for such purpose. He advises these drains to be four inches outside, with a clear water-way of two inches; the tubes must be made with wooden pins instead of nails, and with holes in their sides to admit water. But the cost of construction would in many places make these drains more expensive than tiles, though brought from some distance.

Fig. 37 * represents a *Peat-Tile Drain.* The peat is cut from the ground to be drained in a tile shape, with a tool made for the purpose, (which will be found figured in the chapter on Cutting Drains,) and when dried in the sun, is used as other tiles. As clay is generally absent in the vicinity of such soils, ordinary tiles would be expensive. With the above tool a man may cut from 2,000 to 3,000 peat tiles in a day. This plan is applicable only to mossy light soils.

Sod Draining is a very imperfect mode of draining land (the mode of doing the work is described in a subsequent chapter), and cannot well be recommended for adoption upon principle, as it may often fail. But notwithstanding this, it has been practiced in England for clay soils with considerable success. In the Journal of the Royal Agricultural Society for 1842, it is stated by Lord Spencer, who devoted much attention to agricultural pursuits, that he had drained clay land

* See page 111.

in that way in 1814, and that very little of it had since required to be renewed.

Another method of draining is performed on strong clay land by the *Mole-Plough.* Its object is to make a small opening in the soil at a given distance from the

Fig. 27.

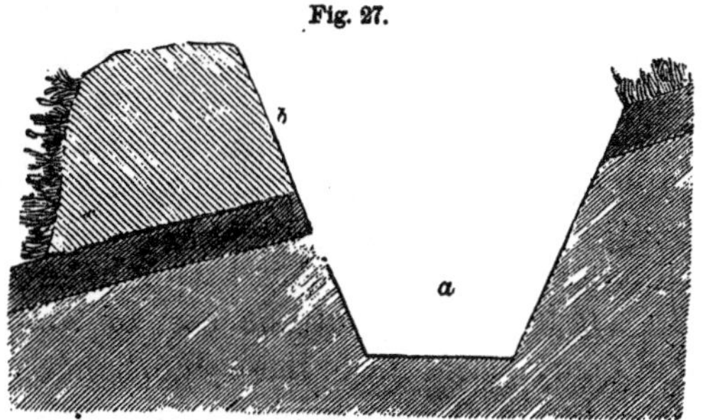

AN OPEN SHEEP DRAIN ON GRASS.

surface, in the form of a mole-run, to act as a duct for the water that may find its way into it. It makes the pipe or opening in the soil by means of an iron-pointed

Fig. 28.

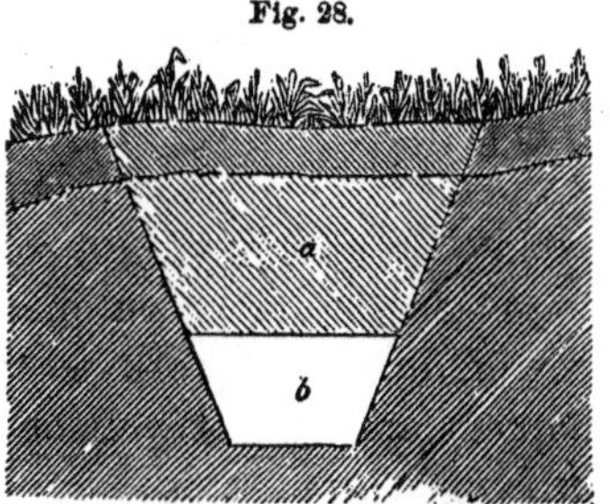

A COVERED SHEEP DRAIN IN GRASS.

cone, drawn through the soil by the application of a force considerably greater than that applied to a common plough.

The mole-plough, Fig. 40, as a draining machine can

never be of much utility except in stiff clay land. In all those subsoils where boulders occur, whether large or small, the mole-plough is inapplicable, its usefulness is limited to such subsoils as consist of pure alluvial clays.

Surface Drains have little variety, being principally open ditches; but it adds much to their permanence and efficiency, particularly where they are made in grass land, if their bottom and sides are protected by stone laid dry, and where stone is plentiful, the expense is not great; and if stone is scarce, slate or a slab of wood with the saw cut upwards, may be substituted for the bottom with stone sides.

The Figures 27 and 28 represent two kinds of drains applicable only to grass lands. They are generally called Sheep Drains. The one is open; the other closed at top, but it has the disadvantage of being liable to be injured by the passage of cattle over it, which would, by their weight, if grazing, tread in the top soil, and so obstruct the drain.

CHAPTER VIII.

.HE MATERIALS ADAPTED TO THE CONSTRUCTION OF DRAINS.

We now come to one of the most important considerations in drainage, *the materials with which to make the drains.* They are numerous, and much will depend in the selection to be made from them, on the question of comparative cost.

The things to be chiefly looked at are, efficiency of operation, durability and cost.

To secure complete efficiency of the drainage when the work is done, should be the first point, and the mode of doing it will be decided upon by reference to the description of soil, the situation, and other particulars, the effects of which have been discussed in the preceding pages. *Permanency and durability* are the next things, and here the same considerations will to some extent operate. In land of great inequality of surface, and which presents in places considerable declivities, the wear and tear of drains will be much greater than in those in which there is but slight inclination in the direction of the drains. Consequently, to secure equally durable work, they must be necessarily constructed of stronger materials in the one case than is required in the other. In all covered drains, the main drains should be very securely made with the best material that the locality affords, or that can be obtained at a cost that is not manifestly disproportioned to the object; because the whole drainage is impeded, or perhaps stopped altogether, if the main drains become defective. Whenever, therefore, the cost is so important a question as to induce the adoption of a less efficient system of drainage than is desirable in a locality, the saving should be effected in the small drains rather than the main, for although a few of the small became inefficient, the injury would be comparatively trifling compared with failure in the main drains.

In reference to the question of the first cost of drainage, there is no doubt *any* drainage, the most temporary or inefficient (comparatively), is better than none. But it is more prudent to drain five acres at a time, and do the work well, and so that it will permanently stand, as to material and workmanship, than to drain four times that number of acres in a temporary bungling way. And this will be cheapest in the end. For the benefit from increase of crops is, in most instances such, as to give greater returns from small pieces of land well drained, than from large ones undrained. Besides which, drainage well executed with the best materials, will remain efficient in most situations for a man's lifetime, and even much longer. There are well-authenticated instances of drains acting perfectly well that had been made one hundred years.

Of the *durability of common brick* when used in drains, there is a remarkable instance mentioned by Mr. George Guthrie, factor to the Earl of Stair, on Culhorn, Wigtonshire. In the execution of modern draining on that estate, some brick drains, on being intersected, emitted water very freely. According to documents which refer to these drains, it appears that they had been formed *upwards of a hundred years ago.* They were found between the vegetable mould and the clay upon which it rested, about thirty-one inches below the surface. They presented two forms—one consisting of two bricks set asunder on edge, and the other two laid lengthways across them, leaving between them an opening of four inches square for water, but having no soles. The bricks had not sunk in the least through the sandy clay bottom upon which they rested, as they were three inches broad. The other form was of two bricks laid side by side, as a sole, with two others built on end on each other at both sides, upon the solid ground, and covered with flat stones, the building being packed on each side of the drain with broken bricks.

From the descriptions of different modes of making drains given in a preceding chapter, it will have been seen that the materials required are used for the purpose either of filling or for building ducts in open or closed drains; and for closing up in the latter the cuts in the surface which were made to enable those ducts to be formed.

To *estimate the relative value of different materials,* durability is the most important point. Because, supposing drains to be properly made, the thing wished is, that they should remain in the same state unchanged. *Stones* and *unglazed tiles* are, therefore, the best of all materials. For tiles, if well-burnt, are practically as durable as stones, in the situation in which they are placed for drains. And a combination of those two materials, as shown in the Fig. 25, is possibly the very best of all drains for durability and efficiency. Unglazed tiles or pipes alone are now considered, by drainers of experience, to effect perfect drainage, and are usually preferred—even to stones. But if stone drains are built in the manner shown in Figs. 21 and 22, the work being well done, and the stones of the proper size, (which will be presently adverted to,) and well rammed down, except in light soils, they would be equally, if not more permanent, with tiles. In light soils, or if made too near the surface, stone drains are more liable than tiles to be displaced by the action of water, or by deep ploughing; although ploughing can only affect them, if, in making the drains, they are placed too near to the surface.

Tiles for draining are made of various shapes and sizes. One of the best description is the pipe, cylindrical-shaped tile, oval in its bore; and, when used, it is placed with a narrow side downwards. The shape secures the better clearance of the interior from sedimentary accumulation. *Collars*, or short outer tiles to go over the joints, should be used with pipes. Another good drain is *horse-shoe shaped*, and should be placed upon a sole tile, of greater width by three or four inches than its base. This shaped tile is now made with a sole attached, which, for stiff soils, is an improvement; but in light soils it would be more liable to become displaced, than if resting on a separate sole. The following table shows the numbers of tiles required for an acre, of the different lengths made, and placed at the stated distances:

	12 in.	13 in.	14 in.	15 in.	
Drains at 12 feet apart require	3630	3351	3111	2904	per acre.
" 15 " "	2904	2681	2489	2323	"
" 18 " "	2420	2234	2074	1936	"

Drains at	21	feet apart	require	2074	1914	1777	1659	per acre.
"	24	"	"	1815	1675	1556	1452	"
"	27	"	"	1613	1480	1383	1291	"
"	30	"	"	1452	1340	1245	1162	"
"	33	"	"	1320	1218	1131	1056	"
"	36	"	"	1210	1117	1037	968	"

The numbers of each length of tile required at intermediate distances can easily be calculated from these data.

It is of great consequence that drains should, when made, remain free from extraneous substances which would ultimately cause stoppages in them; and as the water in passing to them, holds matter dissolved and in suspension in it, the more perfectly it is filtered before it enters the drain the better. This filtration depends in part on the material, and in part on the mode of making the duct of the drain. And in this respect it is that unglazed tile drains have an advantage over all other materials. For as they are porous, and should be so well formed as to fit close together, the water enters them chiefly through the pores of the tile, and in its passage is freed from all substances not chemically combined with it, except as to so much as enters through the small crevices of the joints between the tiles.

Bricks have been used for draining, and they are nearly if not quite of equal value with stones, for the building of the open duct; but as to filtering, they are less effective than tiles, because they present greater extent of joining surface. They are not likely to be much used, as tiles are more eligible, except for mains in some peculiar situations, where stones would be a cheaper substitute, in general, for tiles.

Slates may be used for the soles of ducts with advantage, and for the same purpose as flat stones in the building of ducts; but their edges being usually less even than tiles, admit dirt more readily at the joints.

When, from any cause, more durable materials cannot be procured, any substance that will keep the drain open for the passage of water may be used; and will of course last proportionably to its perishable or durable nature. The following have been used: *Gravel, scoriæ*, or *clinkers* from furnaces, *sand, leather cuttings, old tanner's bark, brush wood, spray, wood faggots, thorn bushes, wood shavings, peaty earth, turf,* and *straw;* and *light earth* for drains in clay.

Although it is best to avoid the use of wood and other perishable material, yet that it may not discourage those who cannot obtain a better article from any cause, it should be stated that it has been found that drains so made have, in many instances, continued effective for years after the filling substance had rotted away, owing to the earth having solidified sufficiently around the filling before it decayed, to remain open afterwards.

CHAPTER IX.

THE MODE OF CUTTING DRAINS.

THE situation of drains upon the land, will be decided upon from the kind of drainage adopted and the nature of the surface. Before determining on the direction in which the lines of drains should run in the field proposed to be drained, it has been recommended to sink pits here and there, of such dimensions as to allow a man to work in them easily, and to a depth which will secure the exposure of the subjacent strata and the greatest flow of water, the depth varying, perhaps, from five to seven feet. But, driving of lines of drains from the bottom to the top of the field, is the most satisfactory method of obtaining an enlarged view of the disposition of the subjacent strata, and, of course, of the depth to which the drains should be sunk. Such lines of drains will not be useless, they will form the outlets of the system of drains connected with each of them, and for that purpose they should be made in the lowest parts of the field.

The *main drains* for receiving the water from the smaller ones, should first be cut. They will be placed in the lowest parts of the ground. In all cases of covered drains, these mains should have open ducts for the water of some of the kinds described.

Whatever be the system of drainage, of course the

depth of cutting for the general drainage, must be regulated within the limits of the fall that can be obtained for carrying the water off the ground at the outlet.

Whenever the *requisite dimensions of the cuttings* for different kinds of drains are not mentioned in this chapter, they have been given in a preceding one, in which they are figured and explained.

The situation of the main drains being fixed upon, the lines of the drains which run *across* from the mains should be marked off. This can be done by drawing a furrow-slice along each line; but a plan which will not require horses, is to set them off by means of short stakes driven into the ground, or, if the field is in grass, by small holes made in the ground with three or four notches of the spade, and the turf turned over on its grassy face beside each hole.

Suppose that it has been determined to make the drains six feet deep. For this depth, a width at top of thirty inches, and one at bottom of eighteen inches, will be quite sufficient for the purpose of drainage, and for room for men to work in easily; and for a less depth, a less width will suffice. This particular, in regard to the dimensions of the contents of a drain, should always be kept in view when cutting one; as even a small unnecessary addition, either to the depth or width of a deep drain, makes a considerable difference in the quantity of earth to be thrown out, and, of course, in the quantity of stones required for again filling up the excavated space. And provided the parts of a drain are substantially executed, its *width*, beyond that which will secure porosity, cannot render it more efficacious. The rule for the width of a drain is well determined by the ease with which men are able to work at the bottom.

The cutting of the drains is commenced by that of the main drain, which terminates at the outlet; and the operation is commenced at the outlet, or lowest part of the field.

The first operation in breaking ground is to stretch the line for setting off the width of the top of the main drain, and each division thus lined off consists of about twenty-four yards. Three men are the most efficient number for carrying on the most expeditious cutting of drains. While the principal workman is rutting

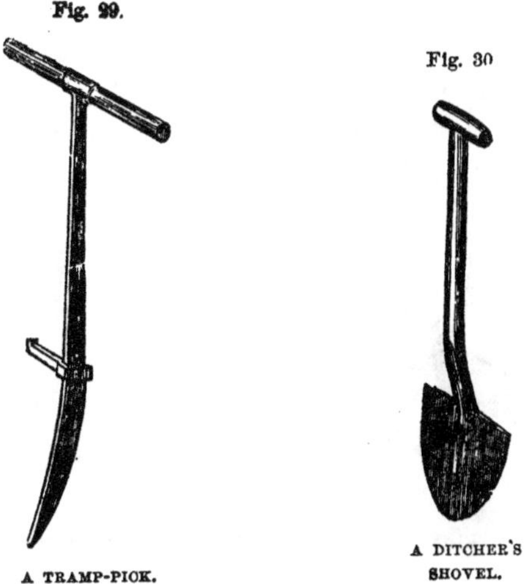

Fig. 29. A TRAMP-PICK.

Fig. 30. A DITCHER'S SHOVEL.

off the second side of the top of the drain with the common spade, the other two begin to dig and shovel out the mould-earth, face to face, throwing it upon the lower side. The first spit of the spade most likely removing all the mould, the first man commences the picking of the subsoil with the foot-pick, Fig. 29; or, if the mould is too deep to be removed by one spit, and requires no picking, the first man digs and shovels out the remainder of it by

Fig. 31.

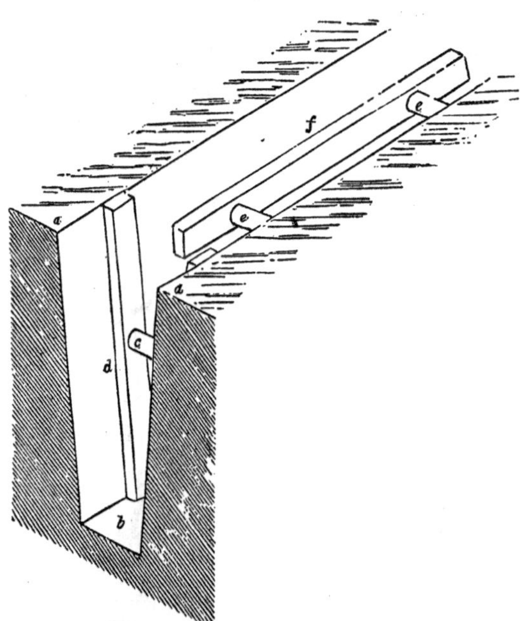

THE POSITION OF PLANKS AND WEDGES TO PREVENT THE SIDES OF DRAINS FALLING IN.

himself with the spade. The mould is thus all removed from the lined-off break or division of the drain. When the picking commences, one man uses the foot-pick, working backward; another follows him with his back with a spade, and digs out the picked earth; while the third comes forward with the shovel, Fig. 30, with his face to the last man, and takes up all the loose earth, and trims the sides of the drain. In this way the first spit of the subsoil is removed. Should the drain prove very wet, and danger be apprehended of the sides falling in, the whole division should be taken out to the bottom without stopping, in order to have the stones laid in it as quickly as possible. Should the earth have a tendency to fall in before the bottom is reached, short, thick planks should be provided, and placed against the loose parts of both sides of the drain, in a perpendicular or horizontal position, according to the form of the loose earth, and there kept firm by short stakes acting as wedges between the planks on both sides of the drain, as represented in Fig. 31, where *a a* are the sides of the drain, *d* planks placed perpendicularly against them, and kept in their places by the short stake or wedge *c*, and where *f* are planks placed horizontally and kept secure by the wedges *e e*.

But if the earth in the drain be moderately dry and firm, another division of four roods may be lined off at top, and the subsoil removed as low as the depth of the former division. Before proceeding, however, to line off a third division, the first division should be cleared out for the builder of the conduit. The object of this plan is to give room to separate the diggers of the earth from the builders of the stones, so as there may be no interference with one another's work, and also to give the advantage of the half-thrown-out earth of the second division as a stage upon which to receive the larger stones, such as the covers of the conduit, and their being easily handed to the builder, as he proceeds in the laying of the conduit in the first division. On throwing out the earth to the bottom of the first break, special care should be taken to clear out the bottom square to the sides, to make its surface even, and to preserve the fall previously determined on.

When a division of the drain has thus been com-

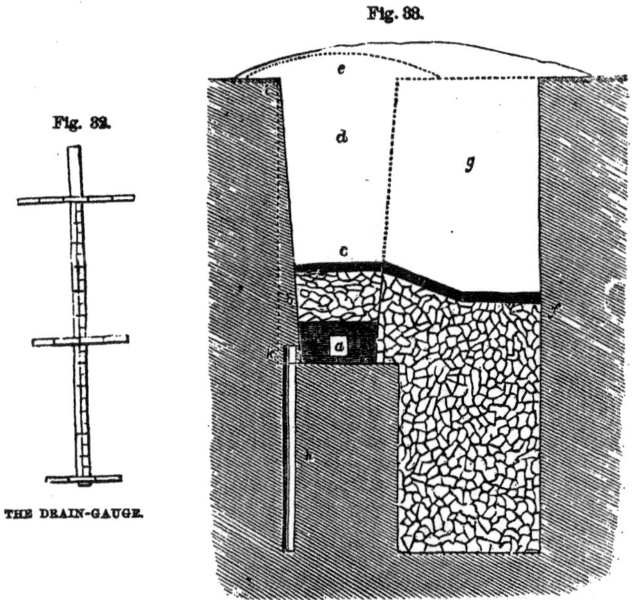

Fig. 32.

THE DRAIN-GAUGE.

Fig. 33.

THE DEEP CONDUITED DRAIN, WITH WELL AND AUGER BORE.

pletely cleared out, it should be ascertained that the dimensions and fall have been preserved correctly, before any of the stones are placed in the bottom.

Instead of measuring the dimensions of the drain with a tape-line or foot-rule, which are both inconvenient for the purpose, a *rod* of the form of Fig. 32

will be found most convenient. The rod, divided into feet and inches, is put down to ascertain the depth of the drain, and then turned partially round while resting on its end on the bottom of the drain, until the ends of its arm touch the earth on both sides. If the arms cannot come round square to the sides of the drain, the drain is narrower than intended; and if they cannot touch both sides, it is wider than necessary.

Should water be supposed or known *to exist* in quantity *below* the reach of a six-feet drain, means should be used to render the drain available for its abstraction, and these means are, *sinking wells and boring* holes into the substrata. A *well* is made as represented by a part of Fig. 33,* where a pit *g* of the requisite depth is cast out on the lower side of the drain *a,* if the ground is not level. A circular or square opening, of three feet in diameter, or three feet in the side, will suffice for a man to work down several feet by the side of the open drain *d;* and, when the stratum which supplies the water is reached, the well should be filled with small stones to about the height of those in the drain, as at *f,* and the whole area of the drain and well, should be covered with dry substances from *f* to *c*, and the earth is filled in again above all, as at *g*. In making such wells, a small scarcement of solid ground, on a level with the bottom of the building of the conduit *a*, should be preserved, so that the building may have a firm foundation to stand upon, and run no risk of being shaken by the operations connected with making the well. Such a well should be sunk *wherever* water

* At *b*, in this figure, is represented the situation for an auger bore into a substratum of porous character.

has been ascertained to be in *quantity* at a lower depth than the drain.

Sub-mains, small, and *cross-drains,* are commenced in the same way as the main drain (but they will not be required to be opened wider at the top than a foot or fourteen inches, for a depth of from three to four feet), namely, by stretching the line, and rutting off the breadth at the top with the common spade. A second man then removes the top-mould with the spade; and if the subsoil is of strong clay, or tiles alone are to be used in filling the drains, he lays the mould on one side of the drain, and the subsoil on the other. In other kinds of soils and subsoils, and where stones are to be used in conjunction with tiles, the separation of the soils is not necessary. A second man follows, and shovels off all the mould, working with his face to the first man. A third man (for the gang or set of drainers should consist of three, for expeditious and clean work) loosens the top of the subsoil with the tramp-pick, Fig. 29, and proceeds backward with the picking, while the other men are removing the mould along the break or division measured off by the line, perhaps sixty or seventy yards. The second man then removes the loosened subsoil with the spade in Fig. 34, which is narrower than the common spade, being six inches wide at the point, digging with his back to the face of the picker—that is, working backward—and the leading man follows with a narrow-pointed shovel, Fig. 30, called the ditcher's or hedger's shovel, with which he trims the sides of the drain, and shovels out the loose part of the subsoil left by the digger.

Should the drain be very wet, owing to a great fall of rain, or the cut draw much water from the porosity of the subsoil, to secure a proper consistence to the drain, it is better to leave off the digging, at this stage of the work, and proceed to set off another length of line at the top; and it would be expedient to remove the top of the whole length of the particular drain *in hand*, to allow the water time to run off, and the sides of the drain to harden, as perseverance in digging would be attended with risk of the sides falling in. This precaution in digging drains is the more necessary to be adopted, in digging narrow shallow drains than deep ones, as planks cannot be used in them to support the falling sides, as in Fig. 31, because the men could not find room in small drains to work *below* the wedges which keep up the planks. *Should the ground be firm*, the digging may properly be proceeded with to the bottom at once. To effect this, the picking is renewed at the lower part of the drain, and another spit of earth thrown out with a still narrower, though of the same form of spade as in the last figure, being only four inches wide at the point. The leading man trims down the sides of the drain with this spade, and pulls out the remaining loose earth toward him with the scoop, such as in Fig. 35; or throws it out with such a scoop as in Fig. 36; and thus finishes the bottom and sides in a neat, even, clean, square, and workmanlike style.

Fig. 34.

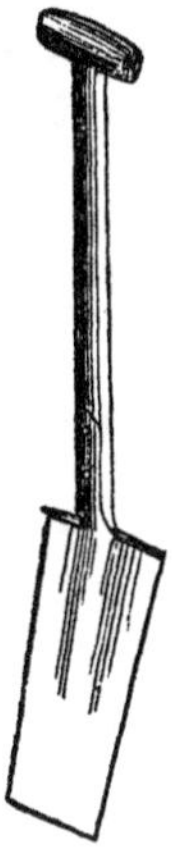

THE NARROW DRAIN SPADE.

What with the experimental cuts, and the first two spits of digging below the mould, the property of the subsoil will be easily determined, and, consequently, *the depth the drain should go*. If the subsoil prove tilly, but still drawing a little water below the mould downward, the drain should be three feet deep, and fifteen inches wide at top; if of intermixed and minute veins of sand, and otherwise of porous materials, then thirty inches of depth will suffice, and twelve inches of width at top; if of quite impervious clay, two feet deep and ten inches width at top will be found sufficient. It is better to cut the drain a little deeper where there is any sudden rise of the surface, and a little shallower where there are any sudden hollows, than to follow the undulations of the ground where these are trifling. As

Fig. 35.

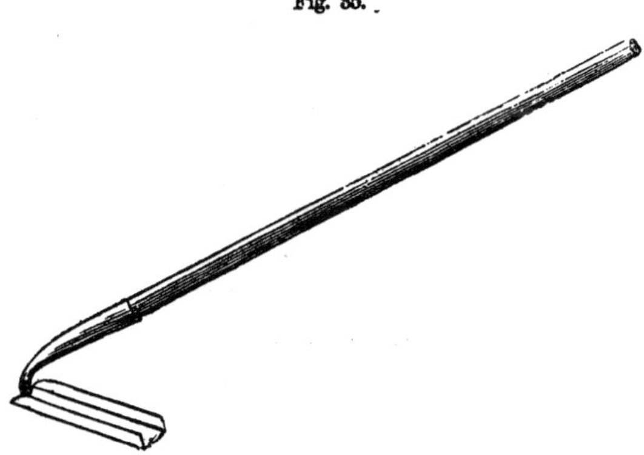

THE EARTH DRAIN SCOOP.

to the distances betwixt the drains in the first case of a tilly but drawing bottom, fifteen feet asunder is wide enough. In the second case of a drawing subsoil, drains at thirty feet asunder will effect as much as in the former case.

The *fall of the ground* can at any time be ascertained by the workmen by a simple contrivance. As the bottom of the drain is cleared out, a damming of four to six inches high will intercept and collect the water seeking its way along the bottom, and by this it can be seen whether the level line of the water cuts the bottom of the drain as far up as it should do according to the specified fall; and a succession of such dammings will preserve the fall all the way up the drain. When the weather is dry, and a sufficiency of water wanting in the drain to adopt this mode of testing the fall, a few

Fig. 86.

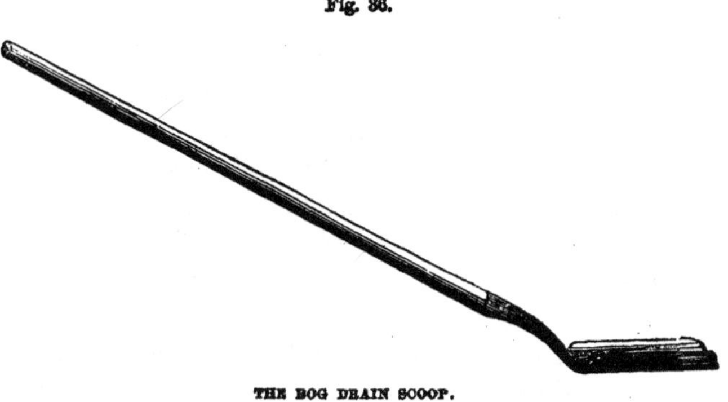

THE BOG DRAIN SCOOP.

pails of water thrown in will detect it, and of course it is only on comparatively level ground that such expedients as these are required.

In all cases of *thorough draining there should be a small drain to connect the tops of the drains* at the upper end of the field, to dry the upper head-ridge, and also to protect the upper ends of the ridges from any oozings of water that might come from the fence ditch, or from any rising ground beyond that end of the field; and it should be of the same depth, though not deeper, than the other drains.

When drains have a course along very long ridges, it is recommended to run a *sub-main* drain in an oblique direction from side to side. *The length of any main drain should not exceed two hundred yards*, without a sub-main drain to assist in carrying off the water; but Mr. Smith says on the subject, that "the practice of throwing in a cross-drain is of no further avail in drying the land, while it increases the length of drain without a proportionate increase of the area drained."* Should the want, however, of proper sized tiles, in any particular part where the quantity of water is greater than over the ordinary surface of the farm, induce the making of a sub-main drain, rather than run the risk of the insufficiency of the drains below, it should be directed across the field, where, if cut of the same depth as the other small drains, those below should be disjoined from it by a narrow strip of ground; but a much better plan is to make the sub-main six inches deeper than the rest of the drains, where it can be so deepened, and it will intercept the water coming from the ground above, while the drains will pass continuously over it.†

In cutting for tile drainage the breadth of the sole tile is the proper width for the bottom of the drain; and except where stones interfere with the digging, the top of the small tile drains need only be opened a few inches wide. Narrow spades with long handles are made for the purpose, which, with the scoop, enables the workman to cut five feet deep with a very small surface cut.

Peat Tile Draining is applicable to the draining of

* Smith's Remarks on Thorough Draining.

† Stephens.

mossy, light soils, where *peat is plentiful.* The peats are made somewhat of the shape of drain tiles but more massive, as may be seen in Fig. 37. They are laid in the drain one *a*, like a tile-sole, and another inverted upon it, as *b*, like a drain-tile, leaving a round opening between them for the passage of the water. These peats are cut out with a spade-tool, Fig. 38. The spade is easily worked, and forms a peat with one cut, without any waste of materials; that is, the exterior semi-circle *b* is cut out of the interior semi-circle of *a*. A man can cut out from 2,000 to 3,000 peats a day with such a spade. The peats are dried in the sun in summer, with their hollow part upon the ground, and are stacked until used; and those used in drains have been found to remain quite hard. The frequent want of clay in upland moory districts renders the manufacture of drain-tiles on the spot impracticable, and their carriage from a distance a serious expense.*

Fig. 37.

b

a

THE PEAT-TILE FOR DRAINS.

Fig. 38.

THE PEAT TILE SPADE.

The cutting of *Sod Drains*, which, it must be remembered, are adapted only to stiff clay soils, is executed by removing the upper turf with the common spade, and laying it aside, for the purpose of making it the wedge at a subsequent part of the operation; and, if the turf

* Quarterly Journal of Agriculture, vol. vii.

is tough, so much the better for the durability of the sod-drain. Another spit is made with the narrow spade, Fig. 34, and the last or undermost one is taken out with the narrowest spade, represented in Fig. 39,

Fig. 39.

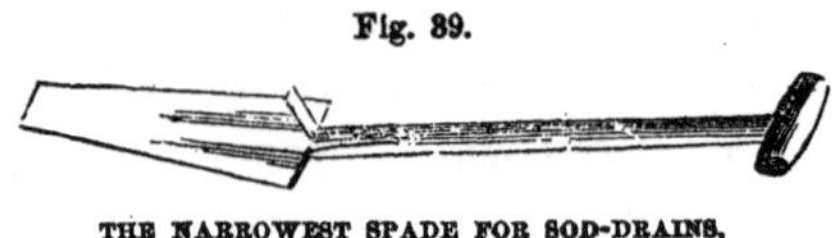

THE NARROWEST SPADE FOR SOD-DRAINS.

which is only two and a half inches wide at the mouth; and, as its entire narrowness cannot allow a man's foot being used upon it in the usual manner, a stud or spur is placed in front at the bottom of the helve, upon which the workman's heel is pressed, and pushes down the spade and cuts out the spit. The depth may be to any desired extent. The upper turf is then put in and trampled or beaten down into the narrow drain, in which it becomes wedged against the small shoulder left on each side of the drain, before it can reach the narrow channel formed by the last-mentioned spade, Fig. 39; and the channel below the turf, being left open, constitutes the duct for the water. It will readily be perceived that this is a temporary form of drain under any circumstances, though it may last some time in grass land, but it seems quite unsuited for arable ground, which is more liable to be affected by dashes of rain than grass land; and, in any situation, the clay in contact with water will run the risk of being so much softened as to endanger the existence of the drain.

Mole Drains are cut at once with the mole draining plough. This plough is of extremely simple construction, as will appear from Fig. 40, which is a view of it in perspective. It consists of a beam of oak or ash wood, six and a half feet in length, and measuring six by five inches from the butt-end forward to four inches square at the bridle *b*. As the beam, when in operation, lies close upon the ground, and is indeed the only means of regulating the depth at which the conduit is to be formed, the lower side is sheathed all over with a plate of iron about half an inch thick. This plate at the proper place, (four ft. four inches or thereabout, from the point of the beam), is perforated for the coulter-box; its fore-end is worked into an eye, which serves as a bridle, and is altogether strongly bolted to the beam. At the distance of a foot behind the coulter-box, a strong stub of wood is mortised into the beam at *c*, standing at the rake and spread which is to be given to the handles. Another plate of iron, of about three feet in length and half an inch thick, is applied on the upper side of the beam; the coulter-box is also formed through this plate, and the hind-part is

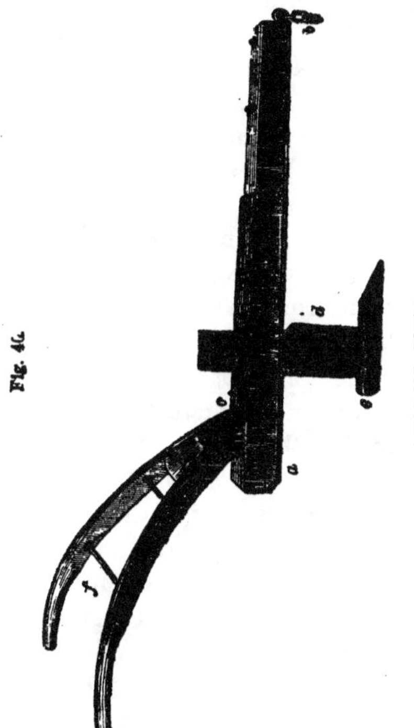

Fig. 40.

THE MOLE-PLOUGH.

kneed at *c*, to fit upon and support the stub, to which, as well as to the beam, the plate is firmly bolted. The two stilts or handles, *c f*, are simply bolted to the stub, which last is of such breadth as to admit of several bolt-holes, by which the *height* of the handles can be adjusted. That which may be termed the head of the plough, is a malleable iron plate, of about two feet in length; that part of it which passes through the beam, and is there fastened by means of wedges, like the common coulter, is seven inches broad and three-quarters of an inch thick. The part *d*, below the beam that performs the operation of a coulter, is nine inches broad, three-quarters of an inch thick in the back edge, and thinned off to a knife-edge in the front. The share or *mole* is a solid of malleable iron, welded or riveted to the head; its length in the sole is about fifteen inches, and in its cross section (which is a triangle with curved sides and considerably blunted on the angles) it measures about three inches broad at the sole, and three and a half inches in height.

In working this plough, the draft-chain is attached to the bridle-eye at *b*, and it is usually drawn by two horses walking in a circular course, giving motion to a portable horse capstan, that is constructed on a small platform movable on low carriage-wheels, and which is moored by anchors at convenient reaches of fifty to sixty yards. The mechanical advantage yielded by the horse capstan, gives out a power of about ten to one, or, deducting friction, equal to a force of about fourteen horses.

When the plough is entered into the soil and moved forward, the broad coulter cuts the soil with its sharp edge, and the sock makes its way through the clay subsoil by compressing it on all sides; and the tenacity of the clay keeps not only the pipe thus formed open, but the slit which is made by the broad coulter permits the water that is in the soil to find its way directly into the pipe. The plough is found to work with the greatest steadiness at fifteen inches below the surface. The upper turf is sometimes laid over beforehand by the common plough, when the mole-plough is made to pass along the bottom of its furrow, and the furrow-slice or turf is again carefully replaced. This is the preferable mode of working this plough, as it serves to preserve the slit made by the coulter longer open than when it terminates at the surface of the turf, where, of course, it is liable to be soon closed up; but the least trouble is incurred when the plough is made to pass through the turf unploughed.

To work the whole apparatus efficiently, two horses and three men are required; they can drain one acre per day. This plough seems fitted for action only in pure clay subsoils, and, when such are found under old grass, it may partially drain the ground with comparative economy; and the process being really economical, it may be repeated in the course of years in the same ground. This mode of draining cannot bear a comparison for efficacy to tile-draining, although it is employed in some parts of England, where its effects are highly spoken of.*

An *improvement in draining with the mole-plough*, has been invented by a Mr. Fowler, who has elevated the beam of the mole-plough on two large wheels behind the coulter, and two small ones in front, and he attaches a rope to the back part of the ploughshare or mole, *c:*

* Stephens.

upon this rope, round pipe tiles, of a foot in length, are strung like beads, close together, and these are carried forward by the rope, and fill the hole made by the share as it advances. *By these means the cut is made and laid with tile by the same operation.* This plough is drawn forward by means of a capstan, as in the case of the ordinary mole-plough. Lengths of rope, of fifty feet each, are successively added as the work proceeds, and the rope is withdrawn at the end of each drain. There is an ingenious contrivance, by which a man, standing on the plough, can, by a wheel and screw, work the coulter up and down, to adapt the depth of the drain to the inequalities in the surface level, his eye being guided by a try-sight on the frame and a staff at the end of the field, set so as to give any inclination to the drain.

After the first cost of the plough, this plan is said to drain at half the ordinary expense. In tolerably free soils, where there is a good fall, it appears unobjectionable; but in stiff clay lands, or in those where the fall is small and the surface level variable, there seems a difficulty in regulating the depth with accuracy enough for such situations.

Mr. Stephens thus describes the cutting of the drain represented in Fig. 41, and which is very imperfect, for many reasons that will be apparent to the reader of the preceding pages, but which may be used, in the opinion of Mr. Stephens, in a *tilly subsoil which draws water* a little, situate in a locality in which *flat stones are plentiful and sufficiently cheap.* Suppose a piece of land, containing two ridges of fifteen feet in width, which had been gathered up from the flat, and in this form of ploughing, there is an open furrow on each side of a ridge. The drains are made in this manner: Gather up the land twice, by splitting out a feering in the crown of each ridge, and do it with a strong furrow. Should the four-horse plough have been used for the purpose, the open furrow will be left sixteen inches wide at the bottom, and if the furrow have been turned over twelve inches in depth, and the furrow-slice laid over at the usual angle of forty-five degrees, the tops of the furrow-slices on the furrow-brow, will be thirty-two inches apart, as from *a* to *a*, Fig. 41. After this ploughing, the spade takes out a trench from the bottom of the open-furrow eight inches wide at top *e*, sixteen

Fig. 41.

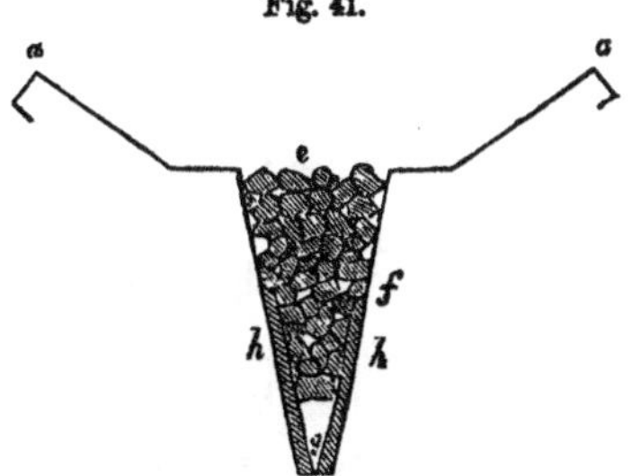

THE FLAT STONE DRAIN.

inches deep by *f*, and three inches wide at bottom at *g*. The depth of the drain will thus be thirty-two inches in all below the crowns of the gathered up ridges. The drain is filled by two stones, *h h*, being set up against its side and meeting in the bottom at *g*; and they are kept asunder by a large stone of any shape, as a wedge, but large enough to be prevented by *h h* descending farther than to leave a conduit *g* for the water. The remainder of the drain is filled to *e* with small riddled stones, with drain-screen, or with clean gravel. The stones are covered over with turf and earth, like any other drain, or with small stones beaten down firmly.

The expense of this method of draining is small: the spade-work may be executed at 1d. the rood of six yards, and of an imperial acre, containing one hundred and sixty-one and a third of such roods, the cutting will cost 13s. 5½d. The flags, at one inch thick and six inches broad, will make fifteen tons per acre, at 4d. the ton, will cost 5s. more. The broken stones, to fill nine cubic feet in the rood of six yards, at 2¼d. per rood, will cost £1 10s. 3½d. more; making in all about £2 8s. 8d.* the acre, exclusive of carriage and ploughing, which, though estimated, will yet make this a cheap mode of draining land so closely as fifteen feet apart. This is of course the cost in Scotland, but it is cheap.

Mr. Stephens adds to the last-described drain the following of a similar character:

"A plan similar may be practiced on *strong clay land.* The open furrow is formed in the same manner with the plough, and, being left sixteen inches in width, the spade work is conducted in this manner: Leave a scarsement of one inch on each side of the open furrow left by the plough, as seen below *a a,* Fig. 42, and cut out the earth fourteen inches wide, perpendicularly, and ten inches deep, as at *b b.* Then cast out from the bottom of this cut, with a spade three or four inches wide, a cut five inches or more in depth *c,* leaving a scarsement of five inches on each side of the bottom of the former cut *b b.* The bottom of the small cut will be found to be thirty-two inches below the crowns of the ridges, when twice gathered up with a strong furrow. The drain is filled up in this way: Take flag-stones of two or three inches in thickness, as *d,* and place them across the opening of *c* upon the five-inch scarsements,

* About $12.

left by the narrow spade; they need not be dressed at the joints, as one stone can overlap the edges of the two adjoining, and they thus form the top of a conduit of pure clay, in which the water may flow. As the water is made to flow immediately upon the clay, it is clear that this form of drain cannot be regarded as a permanent one; though a flag or tile sole laid on the bottom of the cut *c* would render it much more durable. The

Fig. 42.

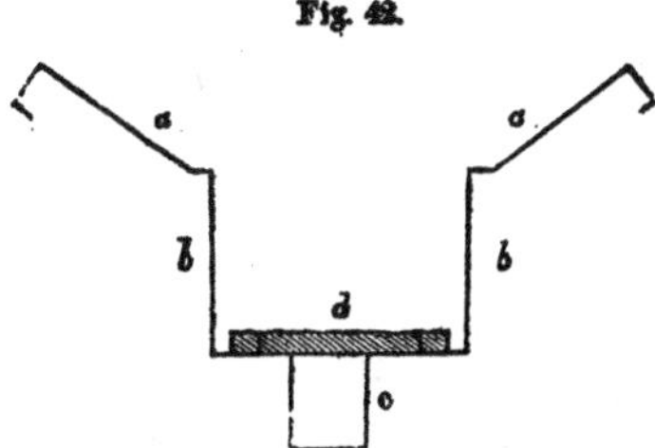

THE CLAY-LAND SHOULDER DRAIN.

cutting of this form of drain, the workmen having to shift from one tool to another, will cost 1½d. the rood of six yards, which, at fifteen feet apart, make 20s. 2d. the acre. The flags for covers will be twelve tons, at 4d. per ton, 4s. more, in all 24s. 2d., but with ten tons of soles, the cost will be 3s. 4d. more, or 27s. 6d.* the acre, exclusive of the carriage of stones and the labor of the plough. After the *joinings* of the flags are covered over with turf, the earth may be returned into the drain with the plough, but with precaution, and probably with the previous assistance of the spade; but, after all, the probability is that *flat* stones cannot be easily obtained in the neighborhood of strong clay, though this *form* of drain may be adopted in any subsoils where flat stones are abundant."

* About $7.

For open surface drains a few directions will suffice. *To cut small drains in grass:* One plan is to turn a furrow-slice down the hill with the plough, and make the furrow afterward smooth and regular with the spade. When the grass is smooth and the soil pretty deep, this is an economical mode of making such drains, which have received the appellation of *sheep drains.* The lines of the drains should all be previously marked off with poles before the plough is used.

A better, though more expensive plan, is to form them altogether with the spade. Let *a*, Fig. 27,* be a cut thrown out by the spade, nine inches wide at bottom, sixteen inches of a slope in the high side, and ten on the low, with a width of twenty inches at top on the slope of the surface of the ground. A large turf *b* is removed by the spade, is laid on its grassy side downward, on the lowest lip of the cut, and the rest of the earth is placed at its back to hold it up in a firm position, the shovellings being thrown over the top to finish the bank in a neat manner.

Another sort of sheep-drain is formed as represented in Fig. 28.* A cut is made six inches wide at bottom, sixteen inches deep, and eighteen inches wide at top. The upper turf *a* is taken out whole across the cut, as deep as the spade can wield it. Two men will take out such a turf better than one. It is laid on its grassy face upon the higher side of the drain, and the earth pared away from the other side with the spade, leaving the turf of a trapezoidal shape. While one man is doing this, the other is casting out with a narrow spade the bottom of the cut *b*. The earth and shovelling should be spread abroad over the grass; and the large turf *a* then replaced in its natural position, and tramped

* See page 91.

down, thereby leaving an open space *b* below it for the water to pass along. This is not so permanent a form of sheep drain as the last, nor is it at all suitable in

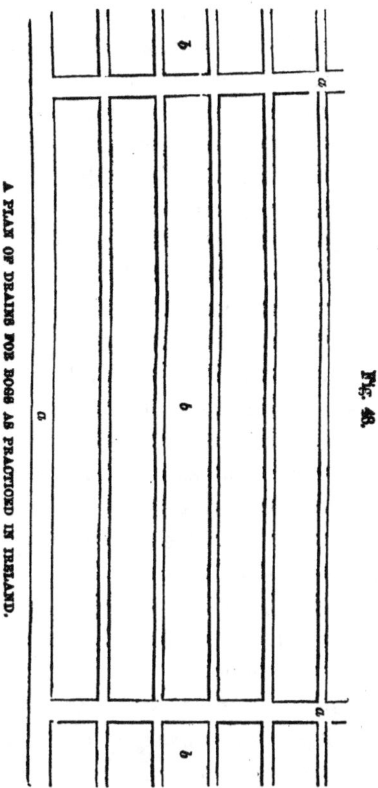

Fig. 48.

A PLAN OF DRAINS FOR BOGS AS PRACTICED IN IRELAND.

pasture where cattle graze, as they would inevitably trample down the turf to the bottom of the drain. It is also a temptation for moles to run along; and when any obstruction is occasioned by them or any other burrowing animal, the part obstructed cannot be seen until the water overflows the lower side of the drain, when the turfs have again to be taken up, and the obstruction removed. It forms, however, a neat drain, and possesses the advantage of retaining the surface whole where sheep alone are grazed. Figs. 27 and 28 are drawn on a scale of one-eighth inch to two inches.

Open Surface Drainage has been extensively and successfully applied to *drying bogs* in Ireland. The plan consists of dividing the bog into divisions of sixty yards in breadth, by open ditches of four feet in depth and four feet wide at top, allowance being made for the sliding in of the sides and subsidence by drying, and which movements have the effect of considerably diminishing the size of the drains; and these ditches are connected by parallel drains at right angles three feet three inches in depth, and eighteen inches in width. Fig. 43 is a plan of these drains, where *a* are the large ditches and *b* the small drains. The ditch *a* at the bottom is that which takes away all the water to some large ditch, river, or lake. The fall in the ditches and drains is produced by the natural upheaving of the moss above the level of the circumjacent ground, and, of course, this peculiarity causes all the drainage of the bog to flow toward the land.

The small drains *b*, Fig. 43, are made in this manner: A bog line is stretched at right angles across the division from the large open drain *a* to *a*, sixty yards. The upper rough turf is rutted in a perpendicular direction along the line with a short edging-iron. The line is then shifted eighteen inches, the width of the top of the drain, and another rut is made by the edging-iron. While one man is employed at this, another cuts out a thick turf across the drain with the broad-mouthed shovel, Fig. 44; and, if any inequalities or ruggedness are observed in the wet turf, he makes them smooth and square with a stroke or two with the back of the shovel. The drain is thus left for two months to allow

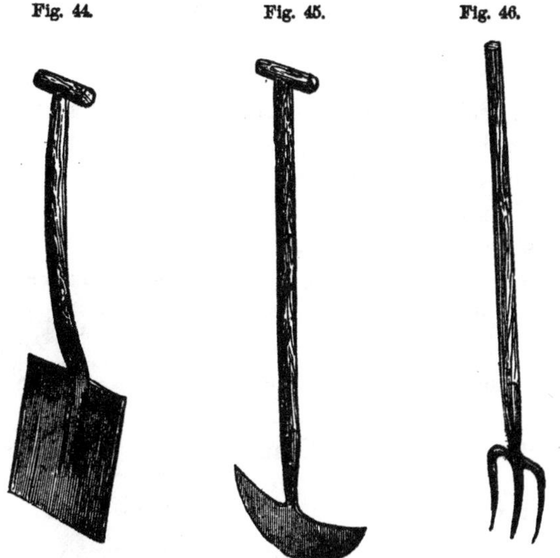

Fig. 44. THE BROAD-MOUTHED SHOVEL. Fig. 45. THE LONG EDGING-IRON. Fig. 46. THE SMALL GRAIP.

the water to run off, the moss to subside, and the turf to dry and harden.

At the end of that time the long edging-iron, Fig. 45, is employed to cut down the sides of a drain in a perpendicular direction two feet three inches (see Fig. 47), and the flat shovel is also again employed to cut the moss into square turfs, which in this case are not thrown out with the shovel—as on account of their wet state they cannot remain on its clear, wet face, when used so far below the hand—but are seized by another man with the small graip, Fig. 46, and thrown on the surface to dry. The work is again left for two months more, to allow time for the water to drain off, and the turfs to dry and harden.

In these four months the moss subsides about one foot. After the two spits of the shovel, the longest edging-iron is again employed to cut down the last spit, which is done by leaving a shoulder, *e e*, five inches broad, on each side of the drain, Fig. 47. The scoop, Fig. 36, is then employed to cut under the last narrow spit, which is removed from its position by the small graip. The scoop pares, dresses and finishes the narrow bottom of the drain, with a few strokes with its back, making the duct *d* one foot deep. As no material but the bog dug out is used to fill these drains, the mode of completing of them had better be here stated. The filling of the drain is performed at this time, and it is done in this manner. The large turf *b*, Fig. 47, which was first taken out, and is now dry, is lifted by the hand and placed, grass side undermost, upon the shoulders *e e* of the drain, and tramped firmly down with the feet. The second large turf *a*, which is not so dry or light as the first, is lifted by the graip and put into the middle of the drain, and the long narrow stripes of turf *c e* separated by the scoop from

Fig. 47.

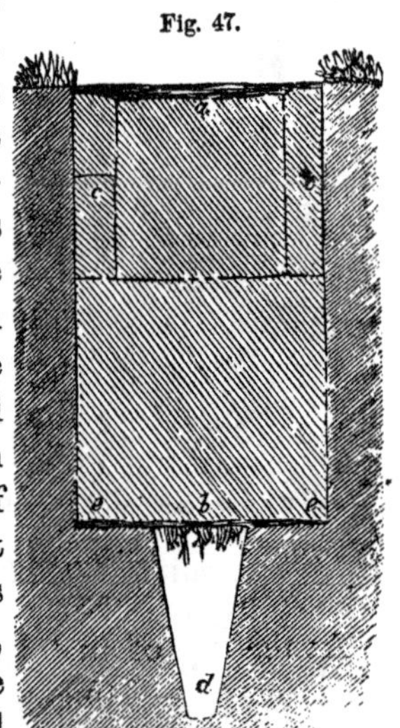

THE SHOULDERED BOG DRAIN.

the bottom, along with other broken pieces, are also placed by the graip along both sides and top of the drain, and all the sods just fill up the subsided drain.

Fig. 47 represents the drain thus finished, which is well suited for the drying of bog, and in its construction possesses the advantage of having all the materials for filling it upon the spot. A bog drain requires no other materials such as wood or tiles, to fill it, there being no material so appropriate or more durable than the moss itself, the slightest subsidence in the drain destroying the continuity of the soles and tiles, whether of wood or clay, while those made of the latter substance will gravitate in the moss by their own weight. The scale of this figure is one-eighth to two inches, or three-quarters of an inch to one foot.*

Boring with the auger may be used in some cases instead of the Well, for the purpose of *piercing through a retentive stratum* into a porous, whereby confined water may be brought up into the bottom of the drain, by altitudinal pressure, and escape; or free water may pass down through the bore, and be absorbed by the porous stratum below. In the first case, the retreat of the water has to be discovered, in making the passage for it to pass away; in the second, it is got rid of by a simple bore. In boring for water at the bottom of a drain, the bore should be made at one side rather than in the middle of the bottom, because any sediment in the water might enter the bore at the latter place and choke it, when the water happened to come up with a small force. In preparation of the bore, let a cut *i k*, Fig. 33,† be made down the side of the drain, and, in-

* Stephens. † See page 104.

serting the auger at *k*, let the bore be made down through the solid ground, in the direction of *b h*, as far as necessary—the orifice of the bore being made at a little higher level than the bottom of the drain, and an opening left in the building there, to permit the water from the bore to flow easily into and join the water of the drain.

Boring-irons may be as useful for finding water for fields, or for draining a bog, or for ascertaining the depth and contents of a moss, or for ordinary draining. The *auger*, *a*, Fig. 48, is from two and a half to three and a half inches in diameter, and about sixteen in length in the shell, the sides of which are brought pretty close together; and it is used for excavating the earth through which it passes, and bringing it up. When more indurated substances than earth are met with, such as hardened gravel, or thin, soft rock, a *punch b* is used instead, to penetrate into and make an opening for the auger. When rock intervenes, then the chisel or jumper *c* must be used to cut through it; and its face should be of greater breadth than the diameter of the auger used. There are *rods* of iron *d*, each three feet long, one inch square iron, unless at the joints, where they are one and a half inch and round, with a male screw at one end and a female at the other, for screwing into either of the instruments, or into one another, to allow them to descend as far as requisite. The short iron *key e* is used for screwing and unscrewing the rods and instruments when required. A *cross-handle* of wood *f*, having a piece of rod attached to it, with a screw to fasten it to the top of the uppermost rod, is used for the purpose of wrenching round the rods and auger, when the latter only is used, or for lifting up and letting fall the rods and jumper or punch, when they are used. The long iron key *g* is used to support the rods and instruments as they are let down and taken up, while the rods are screwed on or off with the short key *e*. Three men are as many as can conveniently

Fig. 48.

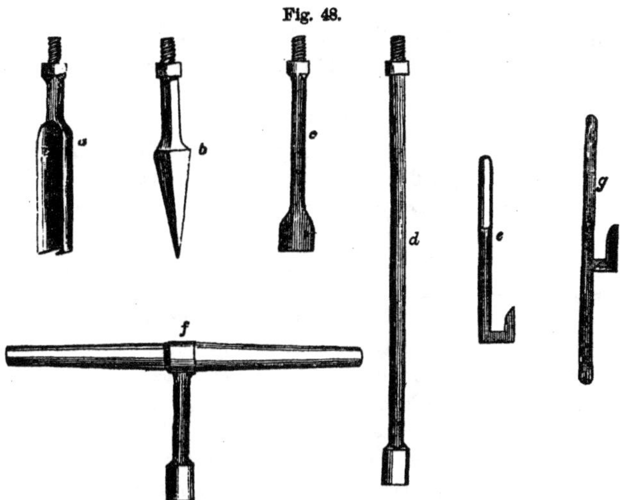

THE INSTRUMENTS FOR BORING THE SUBSTRATA OF DEEP DRAINS.

work at the operation of boring drains. Two men stand above, one on each side of the drain, who turn the auger round by means of the wooden handle; and, when the auger is full of earth they draw it out, and the man in the bottom of the drain clears out the earth, assists in pulling it out, and directing it into the hole. The workmen should be cautious, in boring, not to go deeper at a time, without drawing, than the exact depth that will fill the shell of the auger; otherwise the earth through which it is boring, after the shell is full, makes it more difficult to pull out. For this purpose, the exact length of the auger should be regularly marked on the rods from the bottom upward. Two flat boards, with a hole cut into the side of one of them, and laid alongside of one another over the drain, in time of boring, are very useful for directing the rods in going down perpendicularly, for keeping them steady in boring, and for the men standing on when performing the operation.*

NOTE.—The following is an interesting communication made in 1853 to the Royal Agricultural Society of England, on the subject of "*Swamp Drainage of Lands on Low Levels,* but near rivers or water-courses," by Mr. John Dumolo, of Dunton House, near Colshill, Warwickshire: "I have carried out this method of drainage with the best possible results. In the first place, the water is to be removed from the surface of the land in many places where there is scarcely any fall or outlet, excepting in the adjoining stream, that is to say, when the water in the adjoining river or brook is nearly even with the surface of the land. Now, the drainage of land, under these conditions, in most cases, may be made as effectual as is desirable, and, in many cases, the land be made sufficiently sound for the heaviest of cattle. The method is simple, but may require a little engineering tact to accomplish the object. The drains must be laid even with the bottom of the adjoining river or brook, or at least two or three feet deep in the stream. There is no fear but the water will issue from such drains, and always pass off at a great or greater velocity than those of the stream into which the drainage water will have to enter, by reason that the specific gravity of the drainage water, out of such lands, will, I may say, always be less than the river or brook water is. The only conditions, I would observe, necessary to be stipulated for are: 1st. That a shaft or pipe be fixed at the upper end of the drains, so that the atmospheric pressure may bear thereon, but not allowed to pass through. 2d. That the drains be laid in a proper and judicious manner, with pipes of not less than two inches bore, and that the trenches be well filled

* Stephens.

up. 3d. That the least number of outlets into the discharging stream as practically necessary be made, and that such outlets be at the lowest parts of the stream, as regards the land to be drained. I may remark there will be no detriment to the drainage, should the bed of the drain undulate, or be laid lower than the discharging orifice; but frequently it will be found advantageous to submerge the drains purposely, in order to exclude the atmospheric air, and thus prevent or lessen the danger of stoppages from the sedimentary accumulations of the peroxide of iron, which often abounds in low lands and in bog earths."

6*

CHAPTER X.

ON BUILDING THE CONDUITS OR DUCTS FOR SECURING THE FREE PASSAGE OF WATER THROUGH DRAINS.

In the sectional figures of the different kinds of drains that have been given in a previous chapter, various forms of ducts for drains have been shown, and the materials of which they may be constructed have been pointed out. It only remains in this place, to give such directions as will guide in the building of them.

It is essential that all deep drains, and main drains especially, should be furnished with *built conduits*, that the water may have a free passage in all circumstances, and thereby escape being choked un. which would occasion the expense of relifting and relaying its materials. The relifting of a drain that has *blown*, that is, of one in which the water is forced to the surface of the ground, in consequence of a deposition of mud among the stones preventing its flow under ground, is a dirty and expensive business.

Should the ground be firm, and the drain made in summer, and the length of any particular drain not very great, the *conduit is most uniformly built when begun at the top* and finished at the bottom of the line of drain; but in ground liable to fall down in the sides, or when the drain extends to many roods in length, the safest plan is to build the conduit immediately after the earth is taken out to the bottom.

A very convenient article in the building of conduits in a deep drain is a plank of five inches in breadth, and of from six to nine feet in length, to put down in the middle of the bottom of the drain, to afford a dry and firm footing to the builder, and to answer the purpose, at the same time, of a gauge of the breadth of the conduit, a space of half an inch on each side of the plank giving a breadth of six inches to the conduit. This plank can be easily removed by two short rope-ends, one attached near each end to an iron staple.

Suppose the plank set down at the mouth of the drain in the middle of the cut, *the dyker begins* by leaving a conduit at the mouth of six inches wide, having six inches of breadth of building on each side of it, and six inches high, and using the plank as his foot-board. When the building of these dimensions is finished to the length of the plank, this is carried or pushed by the ropes another length upon the drain, and so on, length after length, until the whole space of drain, when cleared out to the bottom, is built upon. The stones are handed down from the surface to the dyker by the laborer, until the building is finished. The plank is then removed out of the way, the dyker clears the bottom of the conduit of all loose earth, stones, and other matter, with a hand-draw-hoe, five inches wide in the face. Immediately after this, he lays the flat covers, which extend at least three inches on each side over the conduit, they being from two to three inches in thickness; and they lie ready for him on the half cast out division of the drain, from which they are handed to him as he works backward. The open space left between the meetings of the covers, which will not probably have square ends, should be covered with flat stones, and the space from the ends of the covers and flat stones to each side of the drain should be filled up and neatly packed with small stones. In this way the dyker proceeds to finish the conduit in every division of drain. To keep the finished conduit clear of all impediments, the dyker makes a firm wisp of straw large enough to fill the bore of the conduit; and which, while permitting the water to pass through, deprives it of all earthy impurities.

Before the conduit is entirely finished, the drainers throw out the earth of the adjoining division of the drain to the bottom, and the conduit is then built upon it in the same manner as the one just described.

The *openings of main drains at their outlet* should be protected by iron gratings from vermin, which will otherwise enter and cause obstructions.

When the main drain ducts have been built, *the small ducts* may be proceeded with according to whatever mode of making them has been determined upon; in these *the joining of them to the mains* should be carefully built, and it will add to their strength if at each junction an additional thickness of stones or other filling material be placed round the joints to secure them the more effectually from being disturbed or misplaced.

If it should happen that in cutting out the drain, a *quicksand* should be cut into, it will present much difficulty to the builder, and must be met by the best expedients within his command. Planks of wood and props with masonry within them set in cement, is one mode, if the mischief is not too extensive; and another mode that has been found to answer in operation, was this: Thick tough turfs were provided, to lay upon the sand in the bottom of the drain, and upon these were laid flat stones, to form a foundation on which to build a conduit of stones, having an opening of six inches in width and six or seven inches in height. The back of the conduit, when building, was completely packed with turf, to prevent the sand finding its way into it from the sides of the drain; and the packing was continued behind the few small rubble stones that were placed over the cover of the conduit. A thick covering of turf was then laid over the stones, so that the whole stones of the drain were completely encased in turf, before the earth was returned upon them. The filling up was entirely executed with the spade, least the trampling of the horses should have displaced any of the stones.*

In constructing *tile drains* the *cutting* of the main drain should be *entirely finished* before the tiles are laid in it; and immediately after it is finished, it should be measured with the drain-gauge, Fig. 32, to ascertain if it contains the specified dimensions and fall.

While the earth is throwing out toward the narrowest side of the head-ridge, the carts should belaying down the tiles and soles along the open side next the field; they should be placed end to end along the whole line.

The *person* intrusted with the laying of the soles and tiles into the drains, should be one who has been accustomed to that kind of work, and otherwise a good workman; he should be paid by day's wages, that he may have no temptation to execute the work ill.

This person should remain constantly at the bottom of the drains; and, to enable him to do so, he should

* Stephens.

have an *assistant*, to hand him the materials from the ground.

Immediately before proceeding to *lay the sole-tiles*, the man should remove any wet, sludgy matter from the bottom of the drain with a scoop, Fig. 35, and dry earth and small stones can be removed with a narrow draw-hoe, as in Fig. 49, with a two-feet handle *b*, and

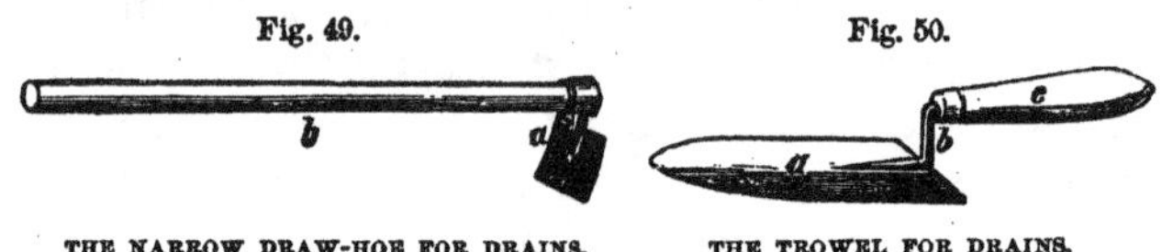

THE NARROW DRAW-HOE FOR DRAINS. THE TROWEL FOR DRAINS.

mouth *a* three inches in width. The sole is firmly laid and imbedded a little into the earth. Should it ride upon any point, such as a small stone or hard lump of earth, that should be removed; and a very convenient instrument for the purpose, and otherwise making the bed for the soles, is a mason's narrow trowel, as in Fig. 50, seven inches long in the blade *a*, nine inches in the handle *c*, and one and a half inches at *b*. After

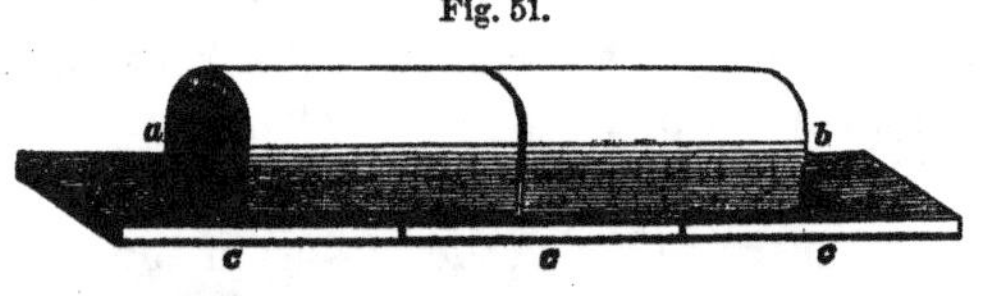

THE DRAIN-TILES PROPERLY SET UPON TILE-SOLES.

laying three soles in length, he examines with a level to see if they are straight in the face, and neither rise nor fall more than the fall of the drain.

After three soles are thus placed, two tiles are set upon them, as represented in Fig. 51, that is, the tiles a and b are so placed as that their joinings shall meet on the intermediate spaces *between* the joinings of the soles c, and this is done for the obvious reason that, should any commotion disturb one of the soles, neither of the tiles, partially standing upon it, should be disturbed. In ordinary cases of water in a *main* drain, a tile of four inches wide and five inches high inside is a good size; and from this size they vary to five and three-quarters inches in width and six and a half inches in height.*

If sole tiles have not been provided, a wooden slab had better be used in the bottom for a sole, rather than lay the drain tiles in the earth. Slates also may be used for the same purpose, and will last longer than wood.

The *covering*, of whatever substance, should be laid in a row or in heaps along the line of the tiles. Turf is the best covering, and it is put over the tiles saddle-wise. If the turf were cut twelve inches broad and eighteen inches in length, it would just lap over the size of tile mentioned above, and rest its end upon the sole on both sides; and, if it be from two to two and a half inches in thickness, the small space left on each side, between the turf and the walls of the drain, would be filled up. When cut, the turfs should be laid one above another, in neat bundles of three or four turfs, which can be easily taken up, and if not used immediately, should be put in large bundles, to keep them supple and moist; but not so kept a long time, in case of their heating and fermenting. If used in summer, in very dry weather, some water should be thrown upon them to keep them moist. A man will cast from four to six cart-loads, of one ton each, per day, accord-

* Stephens.

ing to the smoothness and softness of the ground. Its usual thickness is about three inches, when one square yard will weigh about fifty-four pounds, and of course one ton will cover about forty square yards, or forty roods of six yards with turfs of one foot by one and a half feet.

On being handed to him, the man lays the turf, grass side down, over the tiles in a firm manner, taking care to cause the joinings of the turf to meet as near the middle of the tiles as practicable, and not over the joinings. He takes care not to displace the tiles in the least, when the turf is being put over; and to secure the tiles in their respective places, he puts earth firmly between the covering and the sides of the drain, as high as the turf over the tile. This earth is obtained from the soil that was thrown out; *and if the subsoil is a strong clay, the surface soil is the best*, but a porous subsoil answers the purpose. When all these things have been done, the drain will appear like the small drain, Fig. 53.

The preparations for the junction with the small drains, should be made during the completion of the main drain, in laying which the man should never forget to make the openings at the stated distances the small drains should enter, and, for this purpose, he should be provided with a six-feet rod, marked off in feet and inches, to measure the distances as near as he can, in regard to the fitting of the tiles. The covering of turf should, of course, not be put over the openings left for the small tiles, but the openings should not be left wholly unprotected. A bundle of straw, or rather a turf, until the small drains are connected with them, will be sufficient to protect the openings.

There is a mode of joining tiles in drains that meet one another, that deserves attention. The usual practice is to break a piece off the corner of one or two main-drain tiles, where the tiles of the common drains should be connected with them. Another plan is to set two main-drain tiles so far asunder as the inside width of a common-drain tile, and the opening on the other side of the tiles, if not occupied in the same manner by the tiles of another drain, is filled up with pieces of broken tiles or stones, or any other hard substance. Both plans, however, are highly objectionable, and should never be resorted to where tiles, formed for the purpose of receiving others in their sides, can be procured.

Main Drain Tiles of different lengths should be provided *made with openings* to receive the shouldered end of the furrow-tiles. Fig 52 represents the mode of joining a common drain with a main-drain tile, having an opening in its side. The common tile *b* is not inserted entirely into the main-drain tile *a*, but only placed against it, with a small shoulder, that the openings of both tiles may be always in conjunction.*

Fig. 52.

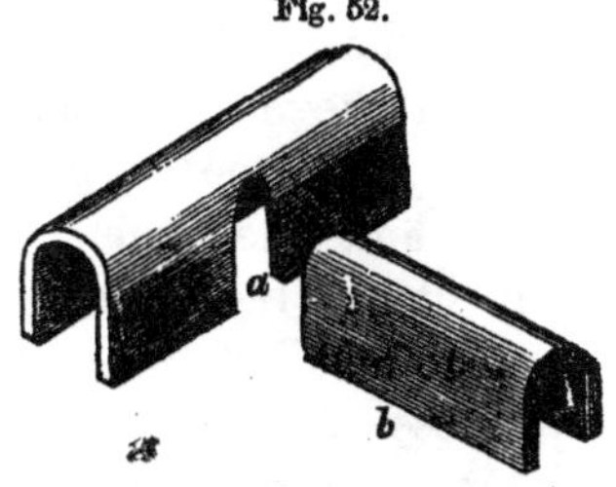

THE JUNCTION OF A COMMON TILE WITH A MAIN-DRAIN ONE.

The *mouth of all main drains* at the outlet, whether in a ditch or river, should be protected with masonry, and dry masonry will do. The last sole which should be of stone, should project as far beyond the mouth as to throw the water either directly upon the bottom, or upon masonry built up by the side of the ditch. The

* Stephens.

masonry should be founded below the bottom of the ditch, and built in a perpendicular recess in its side, with the outer face sloping in a line with the slope of the ditch. The sloping face can be made either straight, which will allow the water to slip down into the ditch, or like steps of a stair, over which the water will descend with broken force. It would be proper to have an iron grating on the end of the outlet, to prevent vermin creeping up the drain; not that they can injure tiles while alive, but in creeping too far up, they may die, and cause for a time a stagnation of water above them in the drain.

The outlet forms the end of the main drain, and its proper place deserves serious consideration. There should be a decided fall from the outlet, whether it is affected by natural or artificial means. If it be very small—and a small fall is all that is absolutely requisite—that is one inch in 150 feet, or three feet in the mile, as indicated by the spirit level—the open ditch into which the main drain issues should be scoured deep enough for the purpose, even for a considerable distance; and when this expedient is adopted, it will be requisite to see every year that the outlet is kept open, and the ditch scoured as often as necessary for the purpose. It is desirable, however, to have a greater fall where the ground admits of it.

If the ground fall uniformly toward the main drain over the whole field, the small drains should be proceeded with immediately after the main drain is finished; but should any hollow ground occur in the field too deep for its waters to find their way direct to the main drain, then a *sub-main drain* should be made along the lowest part of the hollow, to receive all the drainage of the ground around it, in order to transmit it to the main drain. The size of sub-main drains is determined by the extent of drainage they have to effect.

Sub-main drains are made in all respects in the same manner as main drains; but they will most probably have to receive small drains on both sides, on account of the position they may occupy in the area of a field, when they will require just double the number of tiles with openings in the side than the main. In order to avoid the interference of sediment from opposite small drains, these should not enter the sub-main directly opposite to each other, nor should their ends enter at right angles, but at an acute angle.

The sub-main drain should be as far below the small drain as the main itself when it receives the small drains directly, and for the same reasons; and the main should be as far below the sub-main as the latter is below the small drains. The way to effect both these purposes is, to make the main drain deeper after its junction with the sub-main. The main drain being tiled, the small ones may be laid.

Small drains, as well as mains and sub-mains, should be completely cast out, gauged, and examined for the fall, before being filled up; and the materials for doing so should be laid down beside them, as in the case of mains. The tiles for small drains are smaller than for mains and sub-mains, being two or three inches wide and three or four inches high, inside measurement. Soles for small drains are of different breadths, but they should always be two inches wider than the tiles, so that they may rest securely on them without danger of slipping from the subsiding of the soil. The filling of the small drains is conducted in the same manner as the mains and sub-mains, and they are finished as represented by Fig. 53.

While casting out the bottom of the end of each small drain, care should be taken in communicating it with the main or sub-main with which it is to be connected, that no displacement of tiles takes place in either; and when the bottom is cleared out, the turf

Fig. 53.

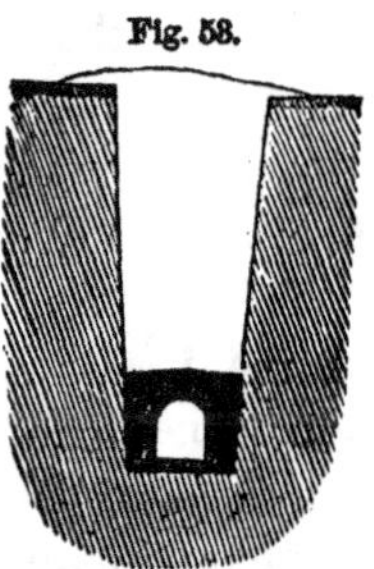

THE SMALL TILE-DRAIN.

or small bundle of straw left in the openings of the sides of the tiles is removed, and the opening examined, and any extraneous matter that may have got into the tiles removed. The places for the entrance of the small drain tiles having been prepared while constructing the main and sub-mains, there will be no difficulty of effecting the junction between the respective sorts of drains. Thus one small drain after another is finished, until the field, having been begun at one side, is furnished with drains by the time the other is reached. The small drain connecting the tops of all the small drains along the upper end of the field, should not be neglected.

When *circular pipe tiles* are used they should be

furnished with "collars" which are short lengths of pipe of diameter sufficient to go over the tiles and cover the joints, a collar being used at each joint. Unless this precaution is adopted these tiles are more liable to become displaced at the joints by the subsiding of the bottom of the drain from wet, than tiles are that are laid upon soles separate from the tiles; because the joints in the latter case are broken with those of the tiles as directed in the instructions for laying them down, and as is seen in the Fig. 51. For the same reason, horse-shoe tiles laid with loose soles are preferable to those made with the soles attached, unless slate or flat stone be first laid in the bottom of the cut of the drain for the tile sole to rest upon, in which case the joints should be broken as above mentioned. But this mode takes more material.

CHAPTER XI.

FILLING UP THE CUTTINGS FOR THE DRAINS, AFTER THE CONDUITS AND DUCTS ARE CONSTRUCTED.

It is of great moment that the filling of the drains after the ducts are built should be done with judgment as to the manner, and with care in executing it.

It is quite a mistake to suppose that all that has to be done is to throw in dirt, stones, or rubbish in any way, so that the cutting is filled and made level with the surface. On the contrary, very much depends upon the proper way of filling, both as to efficiency in the working of the drainage, and also as to durability.

Three principal things in filling must be remembered.

First. That for the drainage to work well in stiff soils, the filling immediately above ducts must in all drains for some distance upward toward the surface, be of a porous open texture, that the water from the neighborhood of them may pass readily into them.

Second. That in deep and thorough drains likewise (*except in clayey* land) the filling *above the last named porous material* should be of a close, tenacious soil, so as to prevent the direct passage of water downward *immediately over* the cutting, for (as has been before explained) that would injure the durability of the drainage by carrying down extraneous matters tending to choke up the drains.

Third. That in thorough drainage of very *stiff clay lands*, the last-mentioned caution must be disregarded, and the drains require to be *filled all the way to the surface with open porous material.*

All *deep* drains, whether furnished with stones or tiles, should receive their supply of water from *below*, and not immediately from above through the soil. Were drains *entirely* filled with loose mould, or other loose materials, the rain, percolating directly through them, will arrive in the drain loaded with as many of the impurities that the soil may contain, as it could carry along with it; and in time they might either collect in quantities in the ducts, or fill up the interstices between the stones. To prevent these mischances, the way is to return the clayey subsoil into the drain, where it will consolidate, and resist the direct gravity of rain.

In the case of *a free open subsoil* to the bottom of the drain, the most retentive portion of the earth may be returned immediately above the *tiles* or *stones* used. *But should the part* of the drain *occupied by the tiles* or stones *be of* strong, *impervious clay*, as much of the loose subsoil should be placed above the tiles or stones as would give an easy access to the water, and all the space above that may consist of clay. "*The general rule, for filling the drains with the earth* that has been thrown out of them is, that, with the exception of strong clay soils—the drains in which should be filled with porous materials, that the water on the surface may descend through them into the duct below—that, with this exception, every kind of drain should be filled near its top with the strongest soil afforded by the drain, in order to *prevent the descent of the water into the drain by the top*, but rather that the water shall seek its way through the ploughed ground, and thence by the porous materials above the duct, and under the clay put in above them into the duct at the bottom. Through such a channel of filtration the water will have every chance of entering the duct in a comparatively pure state.

"*In the case of strong clay soil*, were drains filled up above the tiles with pure clay, the ultimate effect would be that the duct would remain open, but no water would ever enter it. To make them draw at all, there must loose materials be put above the tiles within two or three inches of the plane upon which the sole of the plough moves; and to obtain the greatest depth of loose materials for such drains, they should be made in the open furrows. As they cannot draw but through the loose materials, and are, in fact, covered ditches, they must receive their supply of water like any other ditch, from above; but here the analogy ceases, for instead of receiving their water direct from the top like a ditch, they should receive it by percolation through the ploughed soil, and when the water has descended through the soil, deprived of most of its impurities, it meets the retentive subsoil across the whole area of the ridge, upon which it moves under the arable soil until it meets with the loose materials in the drains, by which it is taken down into the ducts to be conveyed away. The loose materials may be gravel, sand, peaty earth, scoriæ from furnaces, refuse tanners' bark, and such like.

"*In a subsoil that draws only a little water*, were the clayey subsoil returned immediately above the *tiles*, it would have the effect of counteracting the purpose for which the drains were made, because it would curtail the drawing surface to only the height of the tiles themselves. The method, therefore, to fill such drains is to put loose materials immediately above the tiles, to a height not so far as in the case of pure clay drains, but to within one half a foot of the plane of the plough's sole-shoe. Were the drains in such a subsoil, however, filled with *stones*, the case would be different, for these would secure a sufficient drawing surface, and the clayey subsoil may be returned immediately on their top with perfect propriety."*

To proceed with *filling the main Drains*—the duct being built and covered in as before directed†—the stones are thrown in promiscuously upon the covers, until they reach a height of two feet above the bottom of the drain, where they are levelled to a plain surface. They have been recommended to reach the height of four feet, and when the drain is filled with rubble stones entirely, this height is desirable, to give the water plenty of room to find its way into it; but with a conduit such as in Fig. 33, more than two feet seems an unnecessary supply of stones, unless in places where water is more than usually abundant. It has also been recommended to break this upper covering of stones very small; but in deep draining, there seems no good reason for the adoption of such a practice, while it enhances the cost very considerably. Ordinary land stones or quarry rubbish are quite suitable for the purpose, and should any of the stones be unusually large, they can be broken smaller with a sledge-hammer. Should the stones be brought as they are required, the process of filling would be greatly expedited were they

* Stephens.

† See Chapter on Building Ducts.

emptied at once out of the cart into the drain. This could be done by backing the cart to the edge of the drain, and letting the shafts or movable body of the cart rise so gently as to pour out the stones by degrees. To save the edge of the drain, and break the fall of the stones, a strong, broad board should be laid along the side of the drain, with its edge projecting so far as to cause the stones to fall down into the middle of it. A short log of wood placed in front of the board will prevent the wheels of the cart coming farther back than itself. To prevent the stones doing injury to thin covers of drains they should not be allowed to fall direct upon *them*, but *upon the end of the stones previously thrown in*, from which position numbers will roll down of themselves upon the covers without force, and the remainder can be levelled down with the hand before the next cart-load is emptied. There is a very considerable saving in the expense of filling drains in this way, provided it be done in the cautious manner just described, compared with the plan of laying down the stones when the drain is ready to receive them, and then throwing them singly in by the hand.

The levelled surface of the stones should be covered with some *dry material* before the earth is put over them. The best substance for the purpose is turf, and should the field be in grass when it is drained, the turf over the drain could be laid aside, and used for covering the stones. Other materials answer, such as dried leaves, coarse grass, moss, tanners' refuse bark, or straw. The object of placing anything upon the stones is to prevent the loose earth finding its way among them; and, although it is not to be supposed that any of the substances recommended will continue long undecomposed, they, however, preserve their consistence until the earth above them becomes so consolidated as to retain its firmness afterward.

After the drain has been sufficiently filled with stones, the *earth* which was taken out of it should be *returned* as quickly as possible, in case rain fall and wash the earth down its sides among the stones. The filling in of the first earth of a deep drain is usually included in the contract made with the drainer, and done with the spade, because no horse can assist in that operation until the earth has been put in to such a height as to enable him to walk upon it nearly on a level with the ground. The men may either put in all the earth with the spade, or they may put in so much as to allow the plough to do the remainder, but in both cases a little is left elevated immediately over the drain, to subside to the usual level of the ground. There will be much less earth left over the filling than you would imagine from the quantity thrown out at first, and the space occupied by the stones; and it soon consolidates in a drain, especially in rainy weather.

The section of such a drain is seen in Fig. 33*, where *a* is the opening of the conduit six inches square, built with dry masonry, and covered with a flat stone at least two inches thick; and above it is a stratum of loose round stones *b*, sixteen or eighteen inches in thickness. The covering above the stones is *c*, and the earth returned into the drain is *d*, with the portion *e* raised a few inches above the ordinary level of the ground. The *mouths of such conduits when forming outlets*, should

* See Chapter on Cutting Drains.

be protected against the inroads of vermin by *close iron gratings.*

In *filling drains made on the thorough drainage system*, some modification of the proceeding from that described above is necessary. The proper way to commence the filling of the drains in this shallow mode of draining, where they are numerous, is from the upper to the lower end—and not from the lower to the upper—that the bottom of the drain should be cleared out effectually with the scoop before the materials are put in, and this is most easily done down the natural declivity of the ground; and besides, it is at once seen whether the fall has been preserved, by the following of the water down the declivity. In deep-draining the case is otherwise, because in that case the drains being few in number, and each possessing importance, the falls should be previously determined by levelling, and the amount of each levelling marked, by which means they can be preserved as the filling proceeds.

Stones are the most durable material to use *for filling* in the part of the drain immediately above the stone or tile duct, and they must be broken with hammers to the smallness of from two and a half to four inches in diameter. It is a pernicious practice, to mix stones of different sizes in a drain, as such can never assort together, and nothing can be more absurd than to throw in a stone which nearly fills up the bottom of a drain, and is sure to make a dam across it to intercept water. Round small stones, such as are found by the sea-side, are the best for drain filling, because the angular surfaces of broken stones favor the consolidation into a more compact and less porous or open body than round ones: yet as the places that afford small round stones naturally are very limited in number, it is far better to take any sort of quarried stones than leave land undrained, and there is no doubt that almost every sort of stones forms an efficient and durable drain if employed in a proper manner.

Fig. 54.

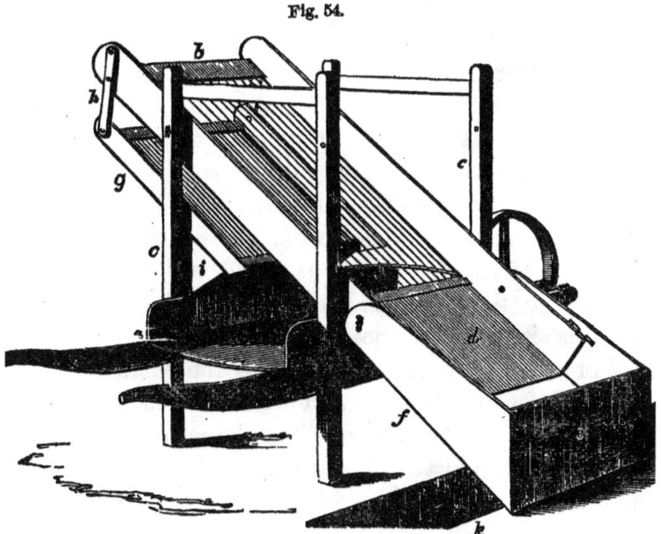

THE DRAIN STONE HARP OR SCREEN.

When these stones are procured, whether in a natural state or broken, they should be screened in order to get them assorted as to size.

The best mode of doing this is by means of a portable *screen* or *harp* for riddling and depositing the stones. Fig. 54, which consists of a wheelbarrow *a*, over which is suspended a screen *b*, having the bars more or less apart, according to the description of materials intended to be used. The upper end is hung upon two posts *c c* about three feet above the barrow; the lower end rests upon the opposite side of the barrow. To this lower end is affixed a spout *d*, attached about ten inches from the lower extremity of which is a board *e*, by means of two arms *f*. Another screen *g*, about one-half the length, and having the bars about half an inch apart, is hung parallel, about ten inches below the larger one. The upper end of *g* is fixed by means of two small iron bars *h* to the upper end of the larger screen; the lower end rests upon a board *i* sloping outward upon the side of the barrow opposite to that on which the spout *d* is situate.

The *stones are put in* in this manner: The earth is all put on one side of the drain. The barrow-screen is placed on the other, so as the board *e*, Fig. 54, attached to the lower end of the spout *d*, shall reach the opposite side of the drain *k*. The cart, with a load of broken stones, is brought to the same side of the drain as the barrow, and the stones are emptied over the top of the screen. The larger ones, rolling down, strike against the board *e*, Fig. 54, and drop into the middle of the drain, without disturbing the earth on either side. The smaller ones, at the same time, pass through the upper screen *b*, and, being separated from the rubbish by falling on the lower screen *g*, roll down into the barrow *a*, while the rubbish descends to the ground on the side of the barrow farthest from the drain.

The best form of shovel for putting the stones over the top of the screen is what is called a frying-pan or lime shovel, represented by Fig. 55.

One man takes charge of the filling of the drain. His duties are to move the barrow forward along its side, as the larger stones are filled to the required height; to level them with the rake, Fig. 56; to shovel the smaller stones from the barrow, spread them regularly over the top of the larger, and beat them down with the beater, Fig. 57, so as to form a close and level surface through which no earth may pass. When the stones are broken in the quarry so as to pass through

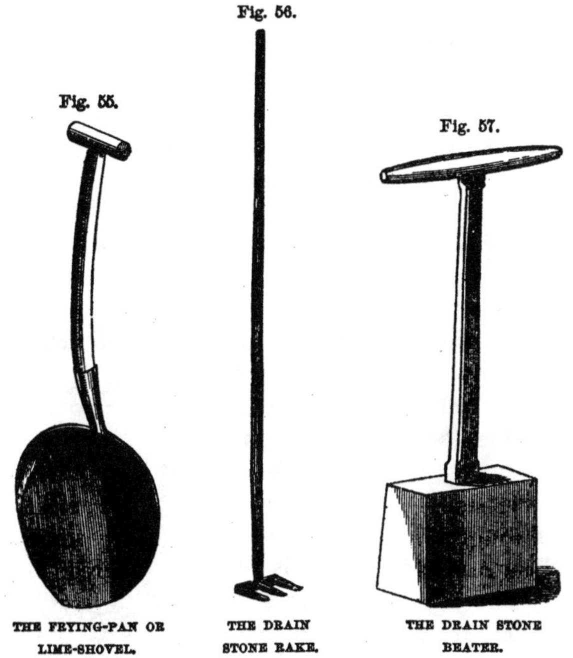

Fig. 55. THE FRYING-PAN OR LIME-SHOVEL.

Fig. 56. THE DRAIN STONE RAKE.

Fig. 57. THE DRAIN STONE BEATER.

a ring four inches in diameter, a quarter of them is so small, or should be made so small as to pass through the wires of the upper screen *b*, Fig. 54, which are one and three-quarters inches apart; and they then will be found sufficient to give the top of the drain a covering of two or three inches deep, which, being beaten closely down, requires neither straw, turf, or anything else to cover them.

With regard to *covering with vegetable substances*, Mr. Roberton says, with much probable truth, that "the only possible use of a covering of straw or turf is to prevent any of the earth when thrown back into the drain, getting down among the stones; but it is evident that such a covering will soon decay, and then it becomes really injurious, because, being lighter (and finer) than the soil, it will, when decomposed, be easily carried down by any water that may fall directly upon the drain; and, if the surface of the stones has been broken so small as to prevent the drain sustaining any injury in this way, then the covering itself must be altogether superfluous."

A drain completed in this manner with stones may be seen in Fig. 58. The dimensions given by Mr. Roberton are thirty-three inches deep, seven inches wide at bottom, and nine inches wide at the height of the stones, which is fifteen inches; and within these dimensions fifteen cubic feet of stones will fill a rood of six yards of drain. The figure represents a drain thirty-six inches deep, nine inches wide at bottom, twelve inches at the top of the stones, and the stones eighteen inches deep. These dimensions give cubical contents of twenty-three and a half feet per rood of six yards; that is, about half as many more stones than the drains of Mr. Roberton.*

Fig. 58.

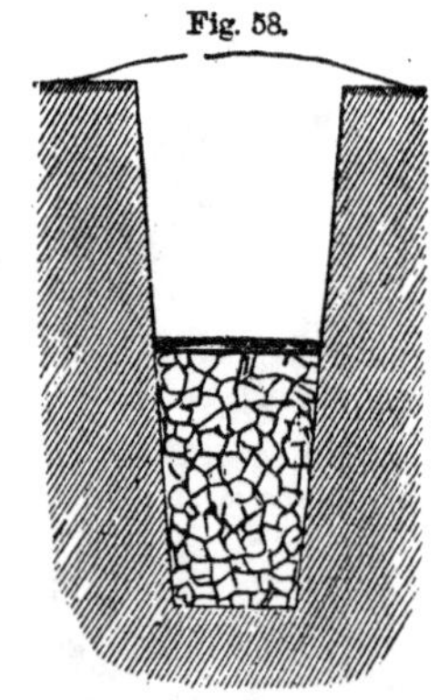

THE SMALL DRAIN FILLED WITH SMALL BROKEN STONES.

* Stephens.

In the foregoing part of this chapter, *directions have been given for filling where stones* are the material used: *the same method will be pursued when* clinkers, broken bricks, coarse gravel, or *any hard material*, is substituted for stones.

For filling bush or brush-wood drains, the mode of placing the wood in them to form the duct, has been already described in Chapter VII., and they must afterwards be filled to the surface on the same plan as if they had stone-ducts and filling; using shaving, spray, wood shavings, tanners' bark, leather cuttings, or chips, or any other porous matter that is available.

Where earth only can be had, sand, peat, and light earth, is the best, if at hand, to place immediately above the duct, in the place of, and in the cases that small stones, when procurable, have above been directed to be used.

When the proper quantity of broken-stones, or other porous material, has been placed in the drains, the next procedure is the *filling up of the drains with the earth that was thrown out of them*, which is returned either with the spade or the plough, or both. When drains are furnished with stones, the plough may be used from the first, giving it as much *land* for the first bout or two as it can work with. If the earth has been thrown out on both sides, a strong furrow on each side of the top of the drain will fill in a considerable quantity of the earth; but, as the earth is generally thrown out on one side of the drain, and the plough can only advance the earth toward the drain while going in one direction, that is, going every other landing *empty*, or without a furrow, a more expeditious mode of levelling the ground, which, in the considerable labor of returning the earth into all the small drains of a field, is a matter of some importance, is to cleave down the mound of earth thrown out, and then take in a breadth of land on both sides of the drain, and gather it up twice or thrice toward the middle of the drain, after which the harrows will make the ground sufficiently level. This species of work, however, is only required when much earth has been thrown out, and thrown a distance from the drain, in deep draining; *but in thorough-draining*, what is accomplished by the plough is done with much less trouble. When the plough alone is used for this purpose, the first two furrows are taken round the mouth of the drain, and fall into it with considerable force; and, where tiles alone are used, such a fall of earth may be apt to break or displace them; and even the steadiest horses, which should only be employed at this work, run the risk of slipping in a hind foot into the drain, which, in attempting to recover, may be overstrained; and such an accident, trifling as it may seem, may be attended with serious injury to the animal. The safest mode, therefore, both for horses and tiles is, in all cases, to put the first portion of earth into the drain with the spade; and there is this advantage attending the use of the spade, that a better choice can be made, if desired, of the earth to be returned. The surface earth may first be put in before the poorer subsoil.*

* Stephens.

CHAPTER XII.

STOPPAGES IN DRAINS.

The stoppage of drains sometimes produces much inconvenience and expense. Stoppages arise principally from two causes. The one, from the ducts of drains becoming displaced in their position, which happens occasionally in stone and in tile drains; and the other, from the drains getting filled up with some extraneous matter which obstructs or wholly prevents the passage of the water, in which case, when they get overcharged, and the force of the water passing through them is considerable, they become "blown" or burst, and the water drives an outlet to the surface.

The first cause generally arises from fault in the original making of the drains, and consequently is much within control; but where very swampy ground or quicksands prevail, additional precaution must be had recourse to in the construction. The second cause of stoppage arises sometimes from bad construction in regard to filling also, but sometimes from the nature of the soil causing the drainage water to hold great quantities of silt and fine sandy particles in suspension; which, unless it be well filtered by the material around the duct of the drain, gets into it, and is deposited in the bottom of the drains in its passage along them. This is to be guarded against to a great extent by care in the filling of the drain, and has been particularly adverted to in the chapter on that subject. But where this deposit takes place, notwithstanding all precaution of that sort that the circumstances within control permit, it is advisable to open the main drains at distances of a few yards, and dig wells at their bottom some three feet or more deep and stone the sides, but leave them empty; so that as the water passes along the drain, it may deposit the suspended matter in them, which it will do; and the situation of these wells or pits being marked, they can be examined and emptied, with little trouble and expense, periodically.

In some localities, minerals and other matters held in solution by water, will, when it is exposed to the greater contact with air in the drains, become disengaged and deposited from the alteration of their chemical combinations. Of this the following is an instance, communicated by a correspondent of the Agricultural Gazette:

"I have lately been much interested by the perusal of the communications of several correspondents on this subject, and as I am a sufferer, though from a different cause to any I have seen stated, I take the liberty of laying my case before you, and shall be much obliged by the suggestion of a remedy. I have on my farm a twenty-acre piece of boggy land, which when I came in possession in 1851, was in a most wretched state, and as full of water as it was possible for land to be. My landlord, Sir W. W. Wynne, Bart., consented to drain it for me, and accordingly Mr. John Green, an experienced person whom he employs, and his staff, were sent over to commence operations. The main drain or outlet was put in 600 or 700 yards in a straight line, with double tiles forming a culvert for the water eleven and a half by five and a half inches, and from six to nine feet deep; from this several other main drains were put in, in other directions, to tap the banks of the elevated ground adjoining, the principal main being of the same sized tiles with others of six-inch and four-inch pipes, with furrow drains of three-inch and two-inch pipes tapping the banks ten, twelve, and in some places sixteen feet deep.

"All went on very satisfactorily, and the undertaking was perfectly successful; the large main culvert discharging at a furious rate, two-thirds full of water, and the land by the spring of 1852 had become quite sound and dry, and continued so until the thaw of the great fall of snow in the beginning of January last, when one day walking across the land, I found that there was a stoppage in one of the drains. I sent over to Mr. Green, but we found the land so charged with water, that the attempt at examination was suspended; but eventually it was found that the large culvert was half filled with a *yellow deposit*, same as enclosed sample, and although perfectly slimy and loose, it had formed in such quantity as to stop the large stream of water in the mains. Orders were given to open holes at twenty yards' distance from each other, a cord was lowered by the stream from one hole to the other, a sweeper was attached to the cord formed of holly or gorse, and by this means, with a deal of trouble, all the main drains were cleaned; the furrow drains were poked with long rods fastened together, and thus all were put in good working order again. The liquor from which the deposit forms, oozes out of the land in many places about one or two feet above the pipes in the drains, and I find that some of the four-inch pipes that were made quite clean about a month ago, have already three-quarters of an inch thick of this yellow substance which I have enclosed, formed and settled in them. It is proposed to make brick-holes similar to pump-wells, say two feet eight inches diameter, frequently along the mains, so as to be able to examine and brush them (as described above) when we find it necessary; but the question, and what we wish to arrive at is, what is the nature of the deposit? Is it likely that it will percolate out of the land (where I presume it has been pent up for very many years) in a year or two, or will it in your opinion continue to harass us for some length of time?"

Upon this letter the editor of the Agricultural Gazette makes the following remarks: "The deposit is no doubt ferruginous, and of the kind indicated by Mr. Parkes in his Newcastle lecture. Rain-water will dissolve oxide of iron in a vegetable soil, but the carbonate of the protoxide of iron, which is thus held in solution, is decomposed on coming in contact with the air in the drain: the iron assumes a higher state of oxidation, which it could not do in the vegetable soil; and the peroxide thus formed, being insoluble in water, is deposited, and fills up the drain. The remedy is by persevering exposure to the air, and by the application of lime, which will engage all the acids likely to dissolve it, and carry it into the drains, to facilitate the perfect oxidation of the mischievous protoxide."*

But one of the most troublesome causes of the stop-

* Agricultural Gazette, April 15, 1854.

page of drains arises from a circumstance not easily guarded against, namely, the disposition of the roots of trees, and of some weeds, to insinuate themselves into drains, where they find the dark but moist situation congenial to a rapid development of their parts. Several remarkable cases of this kind have been recorded, some of which are here extracted, that the importance of taking all practicable precautions against so great an evil may be manifest to the reader, and that he may gather some hints as to the way of remedying it.

Stoppage of drains by the Roots of Trees. "The most remarkable instance which I have ever seen, came under my observation, a few days ago, in the Regent's Park. The drain in question was cut and the pipes laid only about this time last year It was a main drain, formed at the point of obstruction with six-inch pipes. For some considerable distance the side drains, in connection with this main, have a very slight inclination, and, for the last six or eight weeks, the part which these drains traverse (a distance of five hundred or six hundred yards), has indicated a want of activity which could not be satisfactorily accounted for. It occurred to me it was probable that a pipe had given way, and a partial stoppage had, in consequence, taken place, as I could see from the test pot on this main, that the current was not so copious as in another drain which run parallel with it, and received about an equal amount of drainage. I consequently selected two points on the line of the suspected drain, and dug to the pipes in each case. At the one next to the test pot I found the current perfect, but at the other the water rose twelve inches above the pipes. I then proceeded to ascertain the cause, which resulted in the discovery of an immense mass of roots, which occupied the pipes for a continuous line of about six feet. The parent root, where it entered the drain, was only three-eighths of an inch in diameter, and yet the whole volume of pipes nearest to the point of intrusion, was completely filled with the fibrous growth from it, which gradually tapered off to a point. The stoppage had been so progressive, that the water had forced an imperfect course over the pipes, which prevented its rising to the surface, and at once indicating the mischief. The nearest tree was an Italian poplar, standing at the distance of seventy-six feet, and which, no doubt, was the offending party. Here is an example of the necessity for avoiding trees in works of drainage—in a main drain above four feet deep, formed of six-inch pipes, being rendered useless in twelve months by the roots of a tree standing seventy-six feet distant. Of course those trees whose roots have a natural tendency to seek water, and to increase rapidly when in contact with a running stream, are most to be guarded against. The poplars, willows, and alders, are particularly mischievous in this respect, but most trees will produce a great quantity of fibrous roots, when in such a favorable position as is furnished by a rapid current of water in a close drain. This discovery (I allude more particularly to the distance from the tree and rapidity of growth of the roots) gives rise to serious fears as to the durability of draining wherever trees are standing within a distance for their roots to reach the drains. What may be the extent necessary to place drains beyond the reach of free growing trees, I cannot venture to say; but from the rapidity with which, in this instance, so large a pipe has been choked up in a few months, and at a distance of seventy-six feet from the tree, it would appear that drains cannot be safe at a less distance than one hundred feet, and even then it may be doubtful if in time the roots will not reach them at a greater distance.—*P. Mitchell*, 62 Henry street, Portland town, London."*

The Gardeners' Chronicle of the 27th May, 1854, contains an article by the editor, upon the Stoppage of Drains by Roots, in which he says: "A striking example has just occurred at Florence Court, the seat of the Earl of Enniskillen, in consequence of the intrusion of the roots of a large tree of Salix alba, (the white willow,) which, in the short space of eighteen months, did its work most effectually. The pipe tiles, of a four-inch bore, were laid in a drain four feet deep, and covered up in July, 1852; in January of the present year, the drain was observed not to be acting, and was in consequence opened; the pipe was completely choked with roots to the distance of fifty-one feet, and the mass of roots so interwoven as to resemble a thick rope. Two circumstances, more especially, deserve notice in this case. In the first place the pipes were collared, that is to say, each joint in a line of pipes was secured by an external short pipe. Hence it appears that collaring does not offer an effectual barrier to the passage of roots into drains. Secondly, the position and distance of the willow tree in question, seemed to render the introduction of its roots improbable. It stood upon the top of a bank, between six and eight feet above the level of the drained ground, while the

* Agricultural Gazette Dec. 31, 1853.

drain that was choked was sunk four feet below the level, and was nine feet in a direct line from the tree. Nevertheless, the roots, after passing seven or eight feet perpendicularly down the embankment, and then travelling somewhat obliquely nine feet further, reached a depth of four feet, and then having insinuated themselves into some crevice, in eighteen months spread above fifty feet further, filling in their course a four-inch bore as completely as if they had been rammed into the space."

Stoppage of drains by Water Weeds. "My attention was first attracted to this subject thirty-three years since, at Berry Pomeroy. The water that supplies the village is brought down in pipes from a hill about half a mile distant in an easterly direction. These pipes frequently choke; and a force-pump being used, the long hair-like filaments are driven out with violence, and the choking removed. On carefully examining, in a large tub of water, the mass thus extracted, I found it to consist of numerous filaments of great tenuity and of immense length, extending very many yards. There did not appear to be any rootlet, but as far as it was possible to trace the mode of growth, the filaments appeared to spring one from another in parallel lines, and it would seem *to ascend against the stream.* On subjecting the filaments to a powerful microscope, it was discovered that they were copiously in bloom, although quite undiscoverable by the unassisted eye. The blossoms were white, of a pearly lustre, erect, but very minute. It so happens that Denbury, similarly situated on a dry limestone soil, is supplied with water by pipes from a neighboring hill, at nearly the same distance as at Berry Pomeroy. On making inquiries there, it was found that the water coming into the centre of the town, falls into a tank or reservoir, which stores it for the use of the inhabitants: that the pipe conveying the water does not touch the surface of the stored water· and consequently, the filamentous or thread-like weed does not ascend and choke the pipes. This circumstance may be taken advantage of to prevent the nuisance complained of; for if the drain pipes are clean and free, it will suffice to make a perpendicular fall of one and a half or two feet at the nozzle of the flow-pipe, which will prevent the filaments from ascending and choking the drain pipes."—*Agricultural Gazette*, March 18th, 1854.

In the Second Edition of Dr. Lindley's Theory of Horticulture, the following instance is mentioned:—"Patrick Neil mentions an instance of a plant of Ragwort, (Senecio Jacobœa,) which had insinuated the point of its roots into a drain, and had then extended them so much as to fill the drain completely for about twenty feet. And thus it is seen that it is by the point that roots extend, with an indefinite power of branching; and that the finest thread once introduced into a drain pipe, will rapidly become the origin of most extensive mischief, provided the plant is perennial. A still more remarkable case is mentioned in the Gardeners' Chronicle for 1849, of a line of pot pipes from forty to fifty feet long, socketed and cemented, and thought to be perfectly closed, having become so choked by roots as to be unserviceable in fifteen years. In the side of one of the pipes there had been one mere chink, and through that chink some tree had insinuated the point of some root. Once inserted, the point lengthened and divided, and lengthened and divided over and over again; till at last the drain pipe was filled by an entangled mass of fibres, which had pressed so firmly against each other as to form in some places a tolerably perfect mould of the cavity."

Stoppage of drains by roots of Mangold Wurzel. "It is impossible to over estimate the importance of determining the conditions under which drains are most likely to be obstructed by roots, whether of trees, root, or corn crops. My own experience tends to show *that the chief mischief is when there is a perennial flow through* the drain. Some time since, I communicated the fact of Mangold Wurzel roots entirely choking a drain for nearly fifty yards. This drain was cut through the gault clay to its conjunction with the rubble beds of the upper green sand. Tapping a spring, at that point, into this main drain, which was cut directly down the slope, a number of transverse drains are led, cut solely in the gault, and not tapping the spring: consequently, only carrying water after heavy rains, and usually quite dry during the period of vegetation: these drains were not stopped by roots. Now that the question has been taken up, it should be sifted. It is clear that the depth of the drain is no safeguard. It often happens, as in the case before mentioned, when impervious clays underlie pervious strata, under the slope of hills, &c., that open drains might be cut to carry off the perennial water, and that other drains might be so arranged as to carry the water running at intervals. Under all circumstances, I believe that further investigation will prove, that the great mischief by roots is confined to drains carrying perennial water, and that it will be found necessary, as far as may be, to keep such drains distinct from those carrying intermittent water."*

With the view of the ready means of examining and repairing drains when stoppages occur, if for no other reason, *a Map, or plan, should be made at the time of laying drainage, showing the situation of drains*, with their depth in different situations, placed in figures upon it. This will save much labor, time, and expense, on all subsequent occasions when the examination of the drains is requisite.

Stoppage of drains by Weeds. About 1852 or '53, the drains in many different parts of England, that discharged themselves into rivers, were stopped up by a new water weed, to which the name of "*Anacharis Alsinastrum*" has been given. The rapidity of its growth, and the extent of its progress in a short time, over a large portion of the country, has created great interest in it there, but, as we are not aware that it has visited this part of the world, it is not necessary here to pursue the investigation of its habits. Mr. William Marshall, of Ely, has paid much attention to the subject, and has published a small work upon it. Amongst other remedies for its ravages, is one of a singular nature, put forward by the following communication to the *Gardeners' Chronicle* of the 18th November, 1854, from a gardener and nurseryman of respectability, which might be tried upon other vegetable nuisances of an analogous nature: "With reference to the destructive propensities of swans, as directed against the *anacharis alsinastrum*, if you will refer to the *Gardeners' Chronicle* of ten or eleven years back, you will

* Gardeners' Chronicle, June 3, 1854.

find a similar fact registered by myself, which, though it brought me into some little ridicule at the time, was still a fact for all that. I believe, however, that these birds destroy as many weeds, in swimming about and pulling them up, as they do in eating them; and it is quite certain, that almost everything less strong than *Iris Pseud acorus*, must soon perish under their incessant cropping. About four years back, we tried to keep the swans from the water immediately contiguous to the lawn, but the weeds spread rapidly, and we were glad to avail ourselves again of their assistance. Another season, weeds being scarce, the swans attacked a remarkably strong plant (a dense mass, several yards through) of *Nymphœa alba*, and in one night they destroyed almost every leaf, leaving merely the stems and midrib. Such small weeds as the *anacharis* will stand no chance with a few swans, and are sure to be exterminated; and a few geese and ducks will not be found unworthy assistants in the work of destruction.—*William P. Ayres.*"

CHAPTER XIII.

SYPHON DRAINAGE—MR. J. B. DENTON ON THE SUB-SURFACE LINE OF MOISTURE—COST OF DRAINAGE—MISCELLANEOUS MATTERS—TABLE FOR CALCULATING THE CAPACITY OF DRAINS AND DITCHES.

IN the removal of *masses of surface water on a large scale, the principle of the syphon* has been frequently employed with great success. Although connected with the subject on which this work treats, it would too greatly increase its intended limits, to go into that branch of drainage; but the subjoined extract from Dr. Lindley's *Gardeners' Chronicle*, of the 22d October, 1853, is made to call attention to the ingenious mode adopted to relieve the action of the syphon from the accumulative pressure of air. The idea of making the water, as it passed away, the motive power to remove the inconvenience occasioned by its own course through the syphon, although very simple, is, at the same time, not the less interesting.

"The Wigtonshire Free Press, of the 13th September, 1853, describes the successful use of a syphon for the drainage of Culhorn loch, the property of the Earl of Stair. It was found, in the commencement, after working thirty hours, that air had been gradually lodging near the summit, and finally cut off the connection and stopped the discharge. This was calculated upon, and to overcome this evil, two air-pumps, three inches diameter, ten-inch strokes, and twenty strokes per minute, were attached, to draw off the air as it lodged. A small water-wheel, to work these pumps, was placed at the lower end of the syphon, to be driven by the water as discharged. The pumps are connected with the syphon by a one and a quarter inch lead pipe, connected near the summit level at a point where, on an experimental glass model, the air seemed to lodge; and this contrivance has been so far successful. The wheel has gone on working night and day, the pumps drawing air when there is any, or if not, water, till the loch is now lowered nine feet under its former level. It might be drawn lower still, but much difficulty has arisen from the sludge pressing in towards the mouth of the syphon, and from the whole bottom of the loch consisting of a great depth of an impalpable sludge, which must take some time to consolidate and become workable. The syphon referred to is eight hundred and eighty yards, or half a mile long, and is seven inches in diameter. The highest part of the syphon is twenty-one feet above the present surface of the loch, and the longest limb of the syphon is ten feet under the level of the water, giving ten feet of fall. The discharge of the water, at the present time, is about two hundred gallons per minute; but at first, when the loch was at its original height, and the fall greater, of course the discharge was much more. The main part of the syphon consists of cast-iron pipes, five-eighths of an inch thick, with spigot and faucet joints very carefully joined and made air-tight with lead. The contract expense of the iron pipe laid, complete, was 7s. 6d. per yard." (About 180 cents.)

On the *subject of the sub-surface line of moisture, and the deep drainage question*, discussed in the concluding pages of Chapter IV., the following remarks of Mr. J. Bailey Denton, well merit attention:

"If land be drained say four feet deep, gravity takes all water falling on the surface down to the depth of the drains, which establish, as well as maintain in clay soils, a water level at that depth. In porous soils, the inclination of the surface becomes an element in maintaining the water level founded by the drainage. But under all circumstances of soil, it is believed that a line of *moisture*, in deeply-drained land, exists above the line of water level from which the roots of vegetation derive their healthy support of water. Whether this line of moisture really exists as believed, and if it exists, whether it arises from suction, absorption, or capillary attraction, remains for later philosophers to determine. It is a very important question; for if the line of moisture sustained by the water level be identical with the surface of the ground, then evaporation will take effect upon it, and the land remains as cold as ever. That a water level does exist in soils, acting, when near the surface, most prejudicially to healthy vegetation, there can be no doubt, and I will give an instance: It fell to my lot to investigate the valley of the *Test*, for the purpose of drainage, during one of the great windstorms last year, and several trees were blown down. When the base of their roots became exposed to view, they presented as even a surface as the table you have lunched off; and observing that some of the trees had a deeper quantity of earth attached to them than others, I became interested, and found, on examination, that acccording to the height of the ground in which the several trees had grown above a common datum, so was the depth of the soil capsized with the tree. Thus, if a tree which had four feet of soil attached, was compared with the tree which had only three feet, it was found that the ground in which the former tree stood, was exactly one foot higher than the ground of the last."*

Upon the subject of the *cost of draining land*, it is scarcely practicable to give information which, to any useful extent, is of value to the reader in a work of this description. Because so many considerations, of such very different bearing in various localities, enter into the estimate of cost, that one or other of them may cause the most reliable data for one locality to be inapplicable to another. The price of labor, the kind of material employed, the expense of carriage, where, for want of suitable material, it has to be transported, the nature of the soil and subsoil, all influence the question of cost considerably. On the other hand, the kind of material indicated in the preceding pages that is most available, and its value, and the price of labor in his own neighborhood, being well known to every person, it has been thought that the best assistance that can be given to enable the reader to form an estimate of any drainage operations he may contemplate, is to give some information as to the *time in which a certain amount of drainage work has been done*, from which an estimate may be made, making allowance for different soils, of what any given extent of work would occupy, and also some short extracts on the subject of cost from reliable statements by gentlemen who have given the results of their experience.

* Gardeners' Chronicle, August 13, 1853.

From the less labor with which ground is worked in spring, than when hard from the summer's heat, *the cutting of drainage is less expensive in spring than in autumn.* A correspondent of the *Ohio Farmer* states that he cut, in August, one hundred and sixteen rods of drains, eighteen inches deep, and the same width, which cost him, cut with the spade and shovel, nineteen cents a rod, in swamp sour-grass land, counting nothing for boarding or the use of the team. But in March, he cut twenty rods more, in the same way, which cost ten cents a rod only.

The following is an account of the result of a draining match instituted by the Hertfordshire Agricultural Society in England, that took place in July, 1853, which shows *the quantity of work done in a given time:*

"The field selected for the trial was on the margin of the London basin, and showed not only a forbidding subsoil to the drainer, but a badly-farmed and repulsive surface to the visitor, calling out not only for deep draining, but deep cultivation and generous treatment from the tenant. With the stiff yellow clay, was to be observed stones, gravel, and veins of sand, indicating diluvium of the plastic and London clays, in proximity to the chalk on which it rests. The drainers had, therefore, a diversity of soil to work upon, and this fact will explain why some of the shorter were esteemed more meritorious than the longer lengths. There were fifty-one men in the drains, consisting of seventeen gangs of three each. There were eight competitors for pipe laying. *The time devoted to the cutting was five hours.* The cutters worked in gangs of three men each. The first prize was given for a length of one hundred and eight feet, cut four feet deep for two-inch pipes, and the opening at top was twelve inches. The second prize, for eighty-four feet length (stony), with a twelve-inch opening. The third prize for fifty-nine feet length (very stony), with eleven and a quarter to eleven and a half inch opening. The fourth prize for one hundred and sixteen feet length, with thirteen-inch opening."*

The cost of wood drains is alluded to in page 88 of this volume.

In the transactions of the New York State Agricultural Society for 1853, is a statement by a Mr. H. G. Foot, of the expense of draining done upon his farm in Canton, of which he had done about five hundred rods, including open and stone drains. He says one man, in early spring, will throw out from eight to twelve rods a day of ditch, depending upon the depth required; if for covered drain, at a cost of eight cents to one shilling per rod. The drains varied from two to three feet deep, and were filled to one foot of the surface with cobble-stones, covered with pine slabs and turf placed with sod downwards. He considers the expense of filling "somewhat more expensive than the digging," depending partly on the distance the stones have to be drawn.

In draining with tile, in many parts of New York and New Jersey, ten to twelve cents has been about the price paid for cutting drains; and tiles vary in price, according to the size, from ten dollars to twenty dollars per thousand, the cheapest being large enough for the small drains in most situations. The difference of price arises from the size, and from some being made with soles and some without them.

* Gardeners' Chronicle, August 6, 1853.

Cotgreaves Plough for Tile Draining.—In the Transactions of the American Institute of the City of New-York for 1851, is contained the following extract from the London Farmers' Magazine, of a mode of cutting for and laying tiles, which, as the demand for cheap modes of drainage increases, may at no distant period be made available here. It is open to some objections, but where *extensive* works were necessary, these would probably be thought incomparative to its advantages:

"Mr. Cotgreave, of the Rake Farm, near Eccleston, in the neighborhood of Chester, has at length vindicated his county, long stigmatized as the most backward of all English counties in adopting the improvements of the age, in every thing which relates to the amelioration of the soil. By proper drainage, the clay farms will become very productive, and now it can be executed for less than half cost. The Marquis of Westminster, who is extensively engaged in draining his estates, and other eminent agriculturists in the neighborhood, approve Mr. Cotgreave's ingenious invention. Mr. Cotgreave's principle consists of a series of ploughs derived from the *carpenters' plane;* with the exception of the main drains, all the work, even to the obtaining the perfect level of the drain, is performed by the plough plane. Mr. Cotgreave has so adapted his plough that with four horses he can throw out a drain from four to five feet deep. The saving of time is another material object. The work by this process is almost incredibly expeditious, and very little damage is done to the surface; indeed, in grass lands a heavy roller will repair all damages. The cost of workmanship is half the price of manual labor on the present system, and the time occupied *one-tenth*, while the work, to say the very least, is as efficiently and durably performed.

"The working of the plough plane many will doubt. We did so; but we saw and were convinced of its powers and efficiency. All who have witnessed the operation of it are unanimous in their approbation of the plan and their conviction of its full and complete success.

"Ten men and four horses constitute the staff. Without distressing either men or horses, Mr. Cotgreave commences draining two statute acres—4,840 square yards, or 43,560 square feet, each—in the morning, and finally completes, that is, cuts the drains (including the main drain), lays the pipe, fills in and makes good the surface of one statute acre, and half-prepares the second to be ready for work the next day. These plough planes have regulators, which are screws, and by which the plane can be made to shave two, four, five, or six inches thick. When a stone or other obstruction is in the way, the coulter of the plough plane protects the share, and a hooked instrument with a lever is used to extract it before the plough comes back again. One of its great recommendations is, that it is adapted to every variety and condition of soil—can be worked almost independently of the weather unless the ground be too deeply frozen. In fact, those who have witnessed the plough at work are at a loss which most to admire, the absence of complexity in the contrivance or the rapidity and perfect success of the operation. The land owner ought always to have the pipes of the mains and the tributaries on the field ready to be laid down. Cotgreave begins the work with the spade and completes the mains before he begins the tributaries; these he commences by casting out by the plough a sod six inches wide, on the left side of the intended drain over the two acres. This is an immediate service in wet land, for that immediately begins to drain off. He then returns to the first acre and runs his plough plane, casts out the soil and subsoil on the right hand of the drains. In four drafts he cuts down to eighteen inches deep by six inches wide: the first shave being six inches, the last three four inches each. He then takes another plough which cuts six inches deep by two wide; he then lays the pipes by threading them on a half-inch iron bar, one end of which trails in the drain. The rapidity and perfectness with which the pipes are then laid is surprising. A man follows with a sort of paddle, with which he completely adjusts the pipes, and supplies the place of damaged pipes with sound ones. The drain is now ready for filling in, which is rapidly done, and then the whole clod, first turned up in almost one entire piece, is rolled on to its bed. We estimate the benefit of Cotgreave's plan to be a saving of *one-half* the cost, and *nine-tenths* of the time hitherto required.

The cutting of main drains, when done by contract, is sometimes paid for by the running rod of a given width and depth, and sometimes by the quantity of earth removed. In the latter case, as well as for ascertaining the quantity of stones or other filling material, necessary to be prepared, *the cubic contents of cuttings require to be known. A set of Tables applicable to those purposes* is added, which will save much trouble in calculations.

TO MEASURE THE SOLID CONTENTS OF DRAINS OR DITCHES.

Rule.—Take the width at the top and bottom, add these together, their sum divided by 2 will give the mean width; then multiply the width, perpendicular height, and length together; their product, divided by 27, the number of square feet in a cubic yard, (the dimensions being taken in feet and inches,) will give the contents in cubic yards, &c.

Ex.—How many cubic yards are there in a ditch 12 yards long, 7 feet wide at the top, 2 feet at the bottom, and 5 feet deep?

120 yards equal to 360 feet; mean width 4 feet 6 inches.

```
      Feet. In.
       360  0
         4  6
      ---------
      1440  0
       180  0  0
      ------------
      1620
         5
      -----
   27)8100
      -----
       300 cubic yards.   Answer.
```

TO FIND THE CONTENTS FROM THE TABLE.

Find the length in the left hand column, the depth and width in one of the columns on the right of it; then under the depth and width, and opposite to the length, are the contents in cubic yards, &c.

Length.		Depth 2½ feet, mean width 1 ft. 10 in.			Depth 3 feet, mean width 1 ft. 10 in.			Depth 3½ feet, mean width 2 ft.		
Yds.	Ft.	Yds.	Ft.	In. 12th	Yds.	Ft.	In. 12th.	Yds.	Ft.	In. 12th
	1	0	4	7		5	6		7	
	2		9	6		11	0		14	
1			13	9		16	6		21	
2		1	0	6	1	6	0	1	15	
3		1	14	3	1	22	6	2	9	
4		2	1	0	2	12	0	3	3	
5		2	14	9	3	1	6	3	24	
6		3	1	6	3	18	0	4	18	
7		3	15	3	4	7	6	5	12	
8		4	2	0	4	24	0	6	6	
9		4	15	9	5	13	6	7	0	
10		5	2	6	6	3	0	7	21	
20		10	5		12	6	0	15	15	
40		20	10		24	12	0	31	3	
60		30	15		36	18	0	46	18	
80		40	20		48	24	0	62	6	
100		50	25		61	3	0	77	21	
200		101	23		122	6	0	155	15	
400		203	19		244	12	0	311	3	
800		407	11		488	24	0	622	6	
1000		509	7		611	3	0	777	21	

Length.		Depth 4 feet mean width 2 ft.			Depth 4½ feet, mean width 2 ft			Depth 5 feet, mean width 2 ft. 3 in.		
Yds.	Ft.	Yds.	Ft.	In.12th.	Yds.	Ft.	In. 12th.	Yds.	Ft.	In.12th
	1		8			9			11	3
	2		16			18			22	6
1			24		1	0		1	6	9
2		1	21		2	0		2	13	6
3		2	18		3			3	20	3
4		3	15		4			5	0	0
5		4	12		5			6	6	9
6		5	9		6			7	13	6
7		6	6		7			8	20	3
8		7	3		8			10	0	0
9		8	0		9			11	6	9
10		8	24		10			12	13	6
20		17	21		20			25		
40		35	15		40			50		
60		53	9		60			75		
80		71	3		80			100		
100		88	24		100			125		
200		177	21		200			250		
400		355	15		400			500		
800		711	3		800			1000		
1000		888	24		1000			1250		

Length.		Depth 5 feet, mean width 2 feet 6 in.			Depth 4 feet, mean width 3 ft. 3 in.			Depth 4½ feet, mean width 3 ft. 6 in.		
Yds.	Ft.	Yds.	Ft.	In. 12th.	Yds.	Ft.	In. 12th.	Yds.	Ft.	In. 12th
	1		12	6		13			15	9
	2		25	0		26		1	4	6
1		1	10	6	1	12		1	20	3
2		2	21	0	2	24		3	13	6
3		4	4	6	4	9		5	6	9
4		5	15	0	5	21		7	0	0
5		6	25	6	7	6		8	20	3
6		8	9	0	8	18		10	13	6
7		9	19	6	10	3		12	6	9
8		11	3	0	11	15		14	0	0
9		12	13	6	13	0		15	20	3
10		13	24	0	14	12		17	13	6
20		27	21	0	28	24		35	0	0
40		55	15	0	57	21		70	0	0
60		83	9		86	18		105	0	0
80		111	3		115	15		140		
100		138	24		144	12		175		
200		277	21		288	24		350		
400		555	15		577	21		700		
800		1111	3		1155	15		1400		
1000		1388	24		1444	12		1750		

Length.		Depth 4½ feet, mean width 3 ft. 9 in.				Depth 4½ feet, mean width 4 feet.				Depth 4½ feet. mean width 4ft. 6 in.		
Yds.	Ft.	Yds.	Ft.	In.	12th	Yds.	Ft.	In.	12th.	Yds.	Ft.	In. 12th
	1	0	16	10	6		18			0	20	3
	2	1	6	9	0	1	9			1	13	6
1		1	23	7	6	2	0			2	6	9
2		3	20	3	0	4				4	13	6
3		5	16	10	6	6				6	20	3
4		7	13	6	0	8				9	0	0
5		9	10	1	6	10				11	6	9
6		11	6	9	0	12				13	13	6
7		13	3	4	6	14				15	20	3
8		15	0	0	0	16				18	0	0
9		16	23	7	6	18				20	6	9
10		18	20	3	0	20				22	13	6
20		37	13	6	0	40				45	0	0
40		75	0	0		80				90		
60		112	13	6		120				135		
80		150	0	0		160				180		
100		187	13	6		200				225		
200		375	0	0		400				450		
400		750	0	0		800				900		
800		1500	0	0		1600				1800		
1000		1875	0	0		2000				2250		

Length.		Depth 5 feet, mean width 4 feet.			Depth 5 feet, mean width 4 ft. 3 in.			Depth 5 feet, mean width 4ft. 6 in		
Yds.	Ft	Yds.	Ft.	In. 12th.	Yds.	Ft.	In. 12th	Yds.	Ft.	In. 12th
	1	0	20		0	21	3	0	22	6
	2	1	13		1	15	6	1	18	0
1		2	6		2	9	9	2	13	6
2		4	12		4	19	6	5	0	0
3		6	18		7	2	3	7	13	6
4		8	24		9	12	0	10	0	0
5		11	3		11	21	9	12	13	6
6		13	9		14	4	6	15	0	0
7		15	15		16	14	3	17	13	6
8		17	21		18	24	0	20	0	0
9		20	0		21	6	9	22	13	6
10		22	6		23	16	6	25	0	0
20		44	12		47	6	0	50		
40		88	24		94	12	0	100		
60		133	9		141	18	0	150		
80		177	21		188	24	0	200		
100		222	6		236	3	0	250		
200		444	12		472	6	0	500		
400		888	24		944	12	0	1000		
800		1777	21		1888	24	0	2000		
1000		2222	6		2361	3	0	2500		

Length.		Depth 5 feet, mean width 4 ft. 9 in.			Depth 5 feet, mean width 5 feet.			Depth 5½ feet, mean width 4 ft. 6 in.		
Yds.	Ft.	Yds.	Ft.	In. 12th.	Yds.	Ft.	In. 12th.	Yds.	Ft.	In. 12th
	1	0	23	9	0	25	0	0	24	9
	2	1	20	6	1	23		1	22	6
1		2	17	3	2	21		2	20	3
2		5	7	6	5	15		5	13	6
3		7	24	9	8	9		8	6	9
4		10	15	0	11	3		11	0	0
5		13	5	3	13	24		13	20	3
6		15	22	6	16	18		16	13	6
7		18	12	9	19	12		19	6	9
8		21	3	0	22	6		22	0	0
9		23	20	3	25	0		24	20	3
10		26	10	6	27	21		27	13	6
20		52	21	0	55	15		55	0	0
40		105	15	0	111	3		110	0	0
60		158	9	0	166	18		165		
80		211	3		222	6		220		
100		263	24		277	21		275		
200		527	21		555	15		550		
400		1055	15		1111	3		1100		
800		2111	3		2222	6		2200		
1000		2638	24		2777	21		2750		

Length.		Depth 5½ feet, mean width 4 ft. 9 in.				Depth 5½ feet, mean width 5 feet.			Depth 6 feet, mean width 4 feet		
Yds.	Ft.	Yds.	Ft.	In.	12th.	Yds.	Ft.	In. 12th.	Yds.	Ft.	In. 12th
	1	0	26	1	6	1	0	6		24	
	2	1	25	3	0	2	1	0	1	21	
1		2	24	4	6	3	1	6	2	18	
2		5	21	9	0	6	3	0	5	9	
3		8	19	1	6	9	4	6	8	0	
4		11	16	6	0	12	6	0	10	18	
5		14	13	10	6	15	7	6	13	9	
6		17	11	3	0	18	9	0	16	0	
7		20	8	7	6	21	10	6	18	18	
8		23	6	0	0	24	12	0	21	9	
9		26	3	4	6	27	13	6	24	0	
10		29	0	9	0	30	15	0	26	18	
20		58	1	6	0	61	3	0	53	9	
40		116	3	0	0	122	6	0	106	18	
60		174	4	6	0	183	9	0	160	0	
80		232	6	0		244	12		213	9	
100		290	7	6		305	15		266	18	
200		580	15	0		611	3		533	9	
400		1161	3			1222	6		1066	18	
800		2322	6			2444	12		2133	9	
1000		2902	21			3055	15		2666	18	

Length.		Depth 6 feet, mean width 4 ft. 3 in.			Depth 6 feet, mean width 4ft. 6 in.			Depth 6 feet, mean width 4 ft. 9 in		
Yds.	Ft.	Yds.	Ft.	In. 12th	Yds.	Ft.	In. 12th.	Yds.	Ft.	In.12th
	1	0	25	6	1			1	1	6
	2	1	24	0	2			2	3	0
1		2	22	6	3			3	4	6
2		5	18	0	6			6	9	0
3		8	13	6	9			9	13	6
4		11	9	0	12			12	18	0
5		14	4	6	15			15	22	6
6		17	0	0	18			19	0	0
7		19	22	6	21			22	4	6
8		22	18	0	24			25	9	0
9		25	13	6	27			28	13	6
10		28	9	0	30			31	18	0
20		56	18		60			63	9	0
40		113	9		120			126	18	0
60		170	0		180			190	0	0
80		226	18		240			253	9	0
100		283	9		300			316	18	0
200		566	18		600			623	9	0
400		1133	9		1200			1266	18	0
800		2266	18		2400			2533	9	0
1000		2833	9		3000			3166	18	0

Length.		Depth 6 feet, mean width 5 feet.			Depth 6 feet, mean width 5 ft. 3 in.			Depth 6 feet, mean width 5 ft. 6 in.		
Yds.	Ft.	Yds.	Ft.	In. 12th	Yds.	Ft.	In. 12th.	Yds.	Ft.	In. 12th
	1	1	3		1	4	6	1	6	
	2	2	6		2	9	0	2	12	
1		3	9		3	13	6	3	18	
2		6	18		7	0	0	7	9	
3		10	0		10	13	6	11	0	
4		13	9		14	0	0	14	18	
5		16	18		17	13	6	18	9	
6		20	0		21	0	0	22	0	
7		23	9		24	13	6	25	18	
8		26	18		28	0	0	29	9	
9		30	0		31	13	6	33	0	
10		33	9		35	0	0	36	18	
20		66	18		70			73	9	
40		133	9		140			146	18	
60		200	0		210			220	0	
80		266	18		280			293	9	
100		333	9		350			366	18	
200		666	18		700			733	9	
400		1333	9		1400			1466	18	
800		2666	18		2800			2933	9	
1000		3333	9		3500			3666	18	

Length.		Depth 6 feet mean width 5 ft. 9 in.			Depth 6 feet, mean width 6 ft.			
Yds.	Ft	Yds	Ft.	In.12th.	Yds.	Ft.	In. 12th.	
	1	1	7	6	1	9		
	2	2	15	0	2	18		
1		3	22	6	4			
2		7	18	0	8			
3		11	13	6	12			
4		15	9	0	16			
5		19	4	6	20			
6		23	0	0	24			
7		26	22	6	28			
8		30	18	0	32			
9		34	13	6	36			
10		38	9	0	40			
20		76	18	0	80			
40		153	9		160			
60		230	0		240			
80		306	10		320			
100		383	9		400			
200		766	18		800			
400		1533	9		1600			
800		3066	18		3200			
1000		3833	9		4000			

INDEX.

www.ingramcontent.com/pod-product-compliance
Lightning Source LLC
LaVergne TN
LVHW011227110826
845150LV00006B/1565

* 9 7 8 1 4 2 5 5 1 6 1 3 0 *